Monitoring the marine environment

This book is to be returned on
or before the date stamped below

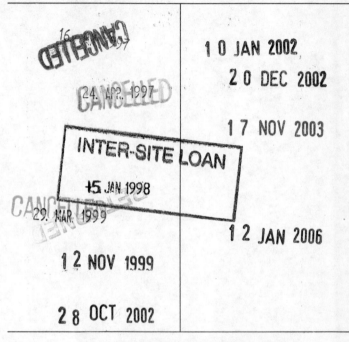

UNIVERSITY OF PLYMOUTH

PLYMOUTH LIBRARY

Tel: (01752) 232323
This book is subject to recall if required by another reader
Books may be renewed by phone
CHARGES WILL BE MADE FOR OVERDUE BOOKS

SYMPOSIA OF THE INSTITUTE OF BIOLOGY NO. 24

Monitoring the marine environment

Proceedings of a Symposium held at the Royal Geographical Society,
London, on 28 and 29 September 1978

Edited by

David Nichols

Department of Biological Sciences,
University of Exeter, England

INSTITUTE OF BIOLOGY
41 Queen's Gate, London SW7 5HU

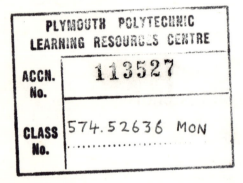
First published 1979
ISBN: 0 900490 12 8
ISSN: 0537–9032

Filmset in 10 on 11pt Times and printed in Great Britain by A. Wheaton & Co. Ltd,
Exeter

Contents

Contributors

Session One
F. G. T. HOLLIDAY (Chairman)   *Nature Conservancy Council, London, England*
W. EIFION JONES, A. FLETCHER, SHEILA J. BENNELL, B. J. McCONNELL, A. V. L. RICHARDS, and S. MACK-SMITH   *University College of North Wales Coastal Surveillance Unit, Marine Science Laboratories, Menai Bridge, Anglesey, Wales*
A. NELSON-SMITH   *Department of Zoology, University College of Swansea, Swansea, Wales*

Session Two
SIR ERIC SMITH (Chairman)   *Formerly Director, The Laboratory, Citadel Hill, Plymouth, Devon, England*
KEITH HISCOCK   *Field Studies Council Oil Pollution Research Unit, Orielton Field Centre, Pembroke, Dyfed, Wales*
GILBERT T. ROWE   *Woods Hole Oceanographic Institution, Massachusetts, USA*
W. F. FARNHAM, E. B. G. JONES, N. A. JEPHSON, and P. W. G. GRAY   *Portsmouth Polytechnic, Portsmouth, Hants, England*

Session Three
G. E. FOGG (Chairman)   *Marine Science Laboratories, Menai Bridge, Anglesey, Wales*
J. M. COLEBROOK   *Institute for Marine Environmental Research, Plymouth, Devon, England*
JAMES M. PARRY and M. A. J. AL-MOSSAWI   *Department of Genetics, University College of Swansea, Swansea, Wales*
RICHARD M. LAWS   *British Antarctic Survey and Sea Mammal Research Unit, Natural Environment Research Council, Madingley Road, Cambridge, England*

Session Four
D. HAMMERTON (Chairman)   *Clyde River Purification Board, Murray Road, East Kilbride, Glasgow, Scotland*
D. H. DALBY   *Botany Department, Imperial College, London, England*
E. B. COWELL and W. J. SYRATT   *British Petroleum Company, Moor Lane, London, England*
N. T. MITCHELL   *Ministry of Agriculture, Fisheries and Food, Fisheries Radiobiological Laboratory, Lowestoft, Suffolk, England*

A. J. NEWTON, A. R. HENDERSON, and P. J. HOLMES
   *Clyde River Purification Board, Murray Road, East Kilbride,
   Glasgow, Scotland*
ROGER MITCHELL   *Nature Conservancy Council, Huntingdon,
   Cambs, England*

Discussion
D. H. ADAMS   *33 Heworth Hall Drive, York, England*
JENIFER M. BAKER   *Orielton Field Centre, Pembroke, Dyfed,
   Wales*
B. CONNELLAN   *7 Spring Hills Tower, Harlow, Essex, England*
T. CROSS   *Department of Zoology, University College, Cork, Ireland*
R. EARLL   *Underwater Conservation Programme, Zoology
   Department, The University, Manchester, England*
J. D. GEORGE   *British Museum (Natural History), Cromwell
   Road, London, England*
P. J. HERRING   *Institute of Oceanographic Sciences, Brook Road,
   Wormley, Surrey, England*
C. R. LAMPLUGH   *Tiverton School, Tiverton, Devon, England*
R. LLOYD   *Ministry of Agriculture, Fisheries and Food, 10
   Whitehall Place, London, England*
PAULINE MARSTRAND   *21 Tyson Road, London, England*
D. C. MONK   *BP Trading Ltd, BP Research Centre, Chertsey
   Road, Sunbury-on-Thames, Middlesex, England*
H. M. PLATT   *British Museum (Natural History), Cromwell Road,
   London, England*
L. REY   *La Jaquiere, CH 1066 Epalinges, Switzerland*
D. RICKARD   *Thames Water Authority, Rivers House, Abbey
   Wood, London, England*
G. A. ROBINSON   *Institute for Marine Environmental Research,
   Prospect Place, The Hoe, Plymouth, Devon, England*
D. L. SIMMS   *Department of the Environment, 2 Marsham Street,
   London, England*
J. H. WICKSTEAD   *The Laboratory, Citadel Hill, Plymouth,
   Devon, England*

Preface

The Symposium reported in this book took as its theme the systematic observation of life in the sea and the effects of human activity on it. The subject was timely for three reasons: firstly, because legislation is now being enforced that designates the obligations of coastal authorities in the matter of human products entering the sea; secondly, because attention is increasingly being focused on the exploitation of living marine resources; and, thirdly, because successful efforts have been made in the last few years to involve interested amateurs in recording aspects of life in the marine environment, thus extending significantly the scope of biological surveillance beyond the resources of most professional scientific teams. The Symposium brought together people interested in a wide range of topics in marine science, and the discussion over such a broad spectrum was especially fruitful.

Thanks are due to those who gave help and advice in putting the programme together; to Mrs Heather Angel and Peter Parks, who provided a static photographic display and film programme respectively; to Mr D. J. B. Copp and his Staff at the Institute of Biology, and also Mrs Alison Leadley-Brown, Symposium Secretary, for practical arrangements; and lastly to Helen Johnson, Publications Officer of the Institute, for preparing proof copies of the papers prior to the meeting and for the production of this book subsequently.

DAVID NICHOLS

Marine monitoring: an introduction

DAVID NICHOLS

*Department of Biological Sciences, University of Exeter,
Hatherly Laboratories, Exeter, Devon, England*

For all its awesome volume, the sea is not a stable system remaining unchanged for long periods of time. On the one hand, there are natural changes responding to climatic and other trends; on the other, there are the effects of human activity and increasing technological stress. Variation of the ocean environment from both agencies will affect the biology of the sea and its ability to yield food, absorb waste, and provide recreation. So the need to take periodic stock of the physical and biological status of various aspects of the marine environment is of great importance if we are to make best use of its bounty. The importance of such monitoring has long been recognized, and various proposals have been put forward as to how international effort might be reappraised and how it could best be harnessed to make optimum use of the resources available to monitor the physical and biological variables in the ocean (see, for instance, Massachusetts Institute of Technology, 1970; UNESCO, 1973; Department of the Environment, 1977).

Previous authors have used the analogy of the 'miner's canary', how it monitors the air and provides an early warning of potential disaster. Some organisms, such as sea-birds, have acted in a similar way, warning of the presence in marine ecosystems of potentially very harmful molecules; yet, unlike the miner, who can reverse out of a danger area when his bird signals a hazard, we on earth cannot escape from our planet if it becomes foul. Neither do we always know what action to take if our 'canary' looks a bit sick.

Monitoring has been defined (MIT, 1970) as follows:

'Systematic observation of parameters related to a specific problem, designed to provide information on the characteristics of the problem and their changes with time.'

It may be easy enough to define, albeit with current jargon, but it is not so easy to advocate a clear-cut programme that makes best use of available resources and expertise, principally because we know all too little about the workings of the whole environmental system, the interactions between atmosphere, ocean, and ecosystem and the inter-relationships of one level in the system with others. What is needed first and foremost is a base-line from which to start, a study of natural ecosystems, so that important changes later that might be the result of harmful activity can be recognized as such and acted upon.

On one scale, this has been initiated in recent years for the $\frac{1}{4}$-million hectares of the British littoral zone, in an attempt to provide a point from which the biological effects of pollution might more easily be recognized. But, as Lewis (1976) has pointed out, some of the present concepts of the permanence of shore ecosystems have far too little regard for the natural processes of variability in the system, such as the coincidence of a number of factors that might permit recruitment of an organism or collection of organisms in an area that has seen no such recruitment for a considerable time, or perhaps a natural decline even to total absence when no spatfall occurs over a period. Sequential monitoring helps in understanding these natural fluctuations so that they can be more easily 'tuned out' when the effect of an artificial agency is being assessed.

A good example of a monitoring exercise of this sort in a fairly discrete marine situation has been reported recently by Tapp *et al.* (1976), in which the area adjacent to the mouth of the River Tees was monitored every three months for a period of six years. Physical, chemical, and biological factors were measured, to determine the magnitude of natural fluctuations in the system and to estimate, so far as possible, the environmental impact of the effluents being brought into the area from the adjacent industrial plants. The authors report very considerable changes in relative abundance of certain species during the period of study, changes that were due to both annual fluctuations associated with reproductive cycles and so on, and longer term variations caused by changes in relative reproductive success and settlement. Yet at the end of the period the investigators were unable to identify any changes attributable to increased contamination by industrial processes. Studies like this one in Teesbay, and indeed like the work around Anglesey and that around a Norwegian oil refinery reported in this volume, are not only a small-scale model of what is required globally, but can also form parts of an international network of linked base-line studies.

Apart from these on-going programmes that help to establish natural fluctuations, there is the urgent need for research into the use of living organisms, either individually or as a population or species, as tools to monitor the quality and state of the environment. This is still a neglected area. Previous work on sensitive organisms has shown that conclusions about their ability as indicators of a change of conditions are all too often equivocal, that individual variation can be extensive, and that the conditions under which a test takes place cannot always be defined accurately enough for the test to have real meaning. Yet 'sentinel' organisms clearly have a most important place in future monitoring work. Stebbing (1976; in press) has overcome some of the problems of individual variation by using a clonal organism, the hydroid *Campanularia*, as the detector: since all the individual colonies used in comparative tests are genetically identical, errors due to individual variation are removed.

For purely practical reasons, it is not surprising that so much effort in monitoring has been directed to shore studies. This volume begins with two aspects of such studies, one (W. E. Jones, A. Fletcher, S. J. Bennell, B. J. McConnell, A. V. L. Richards, and S. Mack-Smith) on mainly base-line work in the region of Anglesey, the other (A. Nelson-Smith) on the recognition of the effects of that ubiquitous shore pollutant, oil. For the past twenty years or so, the sub-littoral has been within fairly easy reach of the modestly equipped biologist, and surveys of shallow-water habitats have gone ahead. One highly encouraging sign of today is that non-biologist divers are keen to take part in these surveys, and recent experience (see, for instance, Nichols, 1978) has suggested that teams of amateur divers can make a real contribution to underwater surveys using SCUBA techniques. The third paper (K. Hiscock) reviews such surveys and outlines a code of practice.

Deeper surveys require more sophisticated apparatus. Much can be done with underwater television and photography, but for some types of work nothing can replace on-the-spot presence of the marine scientist, 'to see things as they are happening', as G. T. Rowe says in his chapter on monitoring with deep submersibles. He describes his own experience in the programme of the DSRV *Alvin* and relates this to future uses of deep-diving vessels in monitoring the remote parts of the ocean.

The next three chapters treat aspects of the biology of the open ocean. First, there is an account (J. M. Colebrook) of monitoring the primary energy source of the ocean, the plankton, mainly from continuous plankton recorders towed behind ships that ply the oceans for purposes other than scientific work. Secondly, there is an account (J. M. Parry and M. A. J. Al-Mossawi) of recent work on the screening systems that have been developed to detect potential mutagenic agents in the seas, and in particular a technique that uses tissues from the common mussel as a concentrator of such substances, and various microbial species as indicators. Thirdly, there is a review (R. M. Laws) of the methods of counting those evocative sea-creatures, the marine mammals. This subject has unusual timeliness, in view of recent action on quotas and other protective measures.

The final section of this book deals with the biological effects of human activity. D. H. Dalby, E. B. Cowell, and W. J. Syratt describe a monitoring programme around a Norwegian oil refinery that has already been underway for six years and is continuing. Then N. T. Mitchell deals with the urgent subject of monitoring radioactivity from coast-sited nuclear power stations. A. J. Newton, A. R. Henderson, and P. J. Holmes give an account of the problems in monitoring the wastes from domestic and industrial sources, as exemplified by the Clyde River Purification Board's responsibilities for disposing of the effluent from half of Scotland's population and industry. Lastly, R. Mitchell describes what steps can be taken to control the effects of human activity on the shores and in the shallow seas around the British Isles, and outlines some strategies that are now being effected to monitor the impact of the activity which could so easily irreparably change marine habitats.

All the monitoring programmes may be said to be working towards one end: prediction. As populations expand and world technology increases, there is a growing need to know *in advance* the environmental consequences of human activities. This requirement, underlined in the developed world by some pretty stiff legislation, is providing an immense challenge to marine biologists, who must strive to understand marine ecosystems, not merely in terms of simple laboratory experiments, but as the integrated and highly complex products of the whole life of the oceans.

References

Department of the Environment (1977) *Monitoring the marine environment of the United Kingdom.* CUEP Pollution Report No. 2. London: HMSO.
Lewis, J. R. (1976) Long term ecological surveillance: practical realities in the rocky littoral. *Oceanography and Marine Biology Annual Review,* 14, 371–390.
Massachusetts Institute of Technology (1970) *Man's impact on the global environment.* Report of the Study of Critical Environmental Problems (SCEP). Assessment and recommendations for action. Cambridge, Mass. and London: MIT Press.
Nichols, D. (1978) Using amateur divers for scientific research. *Australian Federation of Underwater Instructors, Technical Bulletin,* 2, 7–15.
Stebbing, A. R. D. (1976) The effect of low metal levels on a clonal hydroid. *Journal of the Marine Biological Association of the United Kingdom,* 56, 977–994.
Stebbing, A. R. D. (in press) An experimental approach to the determinants of biological water quality. *Philosophical Transactions of the Royal Society,* ser. B.
Tapp, J. F., Taylor, D., Craig, N. C. D., and Lewis, R. E. (1976) *Teesbay marine monitoring: a review, 1970 to 1975.* Brixham: ICI Ltd.
United Nations Educational, Scientific and Cultural Organisation (1973) *Monitoring life in the ocean.* Report of Working Group 29 on monitoring in biological oceanography. UNESCO Technical Papers in Marine Science, 15.

Intertidal surveillance

W. EIFION JONES, A. FLETCHER, SHEILA J. BENNELL,
B. J. McCONNELL, A. V. L. RICHARDS, and S. MACK-SMITH

*University College of North Wales Coastal Surveillance Unit,
Marine Science Laboratories, Menai Bridge, Anglesey, Wales*

Introduction

Surveillance involves keeping a continuous watch on a system and this, if it is to be a useful way of spending one's time, implies that there is a possibility of change. In surveillance of the littoral region, this may mean artificially induced changes in an otherwise unchanging system or natural changes in the system itself. In discussing this, it will be convenient to consider rocky shores first before proceeding to the more complex problems of sand and mud.

Following incidents like the stranding of the *Amoco Cadiz*, the concept of accidental (even catastrophic) change will no doubt seem reasonable but the second may not, at first sight, accord with the ideas on shore ecology which many of us absorbed during our early biological education. In these we were usually introduced to the littoral system as a steady state manifested on rocky shores by a zonation pattern. This pattern was maintained by a harsh environment which, by a series of physical and chemical gradients, imposed conditions which restricted the flora and fauna to a small number of highly specialized species, each with a physiologically limited vertical range. This resulted in a series of identifiable bands of organisms arranged parallel to the sea surface (Stephenson and Stephenson, 1972). A particular shore would have its characteristic zonation pattern and this might be used to help define the conditions, such as exposure to wave action, which characterized the shore (Ballantine, 1961; Lewis, 1964). In broad terms this classical picture is a very sound foundation but, superimposed upon it, the possibility of change has long been recognized. It has been known for a considerable time that a cleared area of littoral rock would go through a sequence of recolonization stages before the climax community was re-established (Evans, 1947). In recent years it has become accepted that the shore communities represent a dynamic balance between a number of organisms, interdependent by competition or predation, and that, as the balance swings under the influence of chance forces, the relative abundance of various species changes. An elegant

1

illustration of this is the demonstration of the interrelationship between *Patella vulgata* L. and other littoral organisms by Lewis and Bowman (1975). Connell (1972) has reviewed this topic and Lewis (1976b) has pointed out the varying importance of physical factors in influencing the biological balance.

Thus we have now an acceptance of community changes determined by biological interactions, moderated by physical (mainly climatic) factors, and varying from shore to shore in the range over which the changes may occur. An understanding of these changes requires long and careful observation and experimental verification. Besides these, however, there are changes which are even more difficult to explain. Reference has been made elsewhere (Jones, 1974) to changes on a rock face at St Anne's Head where a belt of *Pelvetia canaliculata* (L.) Dene et Thur., which had persisted alone above a limpet-grazed face for at least 20 years, was joined in 1965 by a belt of *Fucus spiralis* L. below it and later by the further development of a *Fucus vesiculosus* band below this; with some fluctuations this was the condition in subsequent years. In the same publication, fluctuations in other algal populations were mentioned, including one which has developed further since then: *Nemalion helminthoides* (Vell.) Batt. was collected at several Anglesey localities around 1895 but was absent there in the 1930s, 40s, and 50s. In the early fifties it was present in North Wales only on Bardsey Island. *Nemalion* reappeared on Anglesey in 1962 and flourished for some years until it began to decline in the mid-1970s. In 1977 it was gone and was also found to have disappeared from its old localities on Bardsey Island (Jones, 1978) and to have become scarce on Pembrokeshire shores where it had been common ten years earlier. One may infer that the prime causes of these changes are climatic (though recovery from a local accidental change cannot be ruled out in the St Anne's Head case) and this must raise the speculation that, since some climatic changes are cyclic (Lamb, 1972), the related changes in littoral populations may also be cyclic.

Climatic effects may, however, be quite irregular and unpredictable. One really severe storm may have a far greater effect on the shore than a number of more moderate tempests but the importance of this event may be missed unless the shore is visited soon afterwards so that the effects can be identified. Similarly, both the *Torrey Canyon* incident and the cold winter of 1963 caught the biologists of Britain unprepared (Crisp, 1964) in that we were forced very largely to evaluate the resultant changes by investigations undertaken after the events with incomplete knowledge of the original condition of the affected shores. In fact, it was the realization that another event of this kind would find us equally unprepared which was one of the reasons for undertaking the investigation that forms the main subject of this paper.

A coastal surveillance programme

If we wish to know more about changes on the shore we are faced with the need to collect information over a long period of time. This is bound to be a laborious and, at times, dull occupation and it is worth summarizing the reasoning involved in setting it up.

1. Substratum. Ideally, it would be valuable to collect information systematically from rocky shores and from mobile beaches, both open sandy shores and sheltered muddy flats. There are differences in the magnitude of the tasks, however: rocky shores offer a series of communities which can be inspected without much disturbance and allow accurate numerical assessment *in situ*. Shores of mud and sand, however, have the bulk of their fauna hidden below the surface and involve the digging up of samples which must be evaluated in the laboratory; it is not practicable to carry out this evaluation on the spot, particularly as the impossibility of assessing much of the possible variation from the appearance of the beach means that a random sampling technique must be adopted. This is in itself destructive, and a programme which involved such sampling at, say, monthly intervals on a beach of limited size would have to be questioned on conservation grounds. Thus economy of time and manpower indicated the advisability of concentrating on rocky shores. This is not to say that monitoring the mobile beaches would not be profitable, but fortunately programmes are being undertaken on offshore sediments (such as that conducted by E. I. S. Rees's team in Liverpool Bay) which are producing the kind of information required.

2. Time scale. People who visit the shore only at one time of the year (and, traditionally, there are biologists who do so every Easter but not at other times!) may miss the fact that considerable seasonal changes occur. To establish the extent of the changes, even assuming each year is similar to the next, means a minimum of twelve months' observation. This can be taken as a warning against quick 'base-line' surveys of the kind sometimes undertaken before the start of coastal 'development' works. A survey undertaken in summer is not a sound basis for evaluating the effect of, say, an oil spillage on the site in mid-winter. However, cyclic changes over periods of years can be demonstrated only by sampling over considerably more than the duration of one cycle; a periodicity related to the 11-year sunspot cycle might well require over 20 years of sampling and, if the changes in *Nemalion* abundance mentioned above should be cyclic, we are forced to consider sampling over more than a normal human life span (even in a profession which is well known for longevity!). Long-period sampling raises many problems, such as the continuity of control and interest and, if funds are needed, the reluctance of research councils to support research for longer than the standard three-year period—nor will the prospect attract an academic seeking rapid and frequent publication!

3. Basic strategy. If it is decided to mount a surveillance programme, there arises the question of what exactly should be monitored. There are two possible strategies:

(*a*) The broad-scale approach; this will require a system involving the accumulation of quantitative data on the largest number of species from a series of sites.

(*b*) Accepting that much interdependence exists between species on the shore, it should be possible to reduce the effort required by concentrating on one or a very few dominant species whose fluctuations imply corresponding changes in other species and so provide results which can be extrapolated to give a general picture.

Each scheme has its value. The first will commend itself particularly to the littoral botanist: the interactions between algal species are not well understood and sampling methods are available which can conveniently record data on all the algal species present in a quadrat in one operation. The second is appropriate for the better understood animal inter-relationships (Lewis, 1976a, 1976b) and will commend itself where labour and time are limited or where sampling is proposed over large geographical areas. The two schemes are not basically different—the balance between them will depend on the degree of confidence with which the interdependence of the major species can be predicted.

These considerations were borne in mind when it was decided to set up the University College of North Wales Coastal Surveillance Unit. The reasoning set out below was accepted by Shell UK Ltd, whose generous and enlightened support has enabled the project to continue to the present time. It was decided to concentrate on rocky shores and, in designing the sampling programme, to adopt the following principles:

(*a*) Methods of sampling were to be tested and adopted if they gave a good quantitative representation of the flora and fauna. The number of species sampled should be as large as possible and should include all sedentary species.

(*b*) The unit sampled should be the individual species and the recorded data should retain this information rather than contracting it into an index or other aggregated form.

(*c*) Sampling methods should be objective, repeatable, and standardized so that different workers could produce similar results for the same test quadrats.

(*d*) The sites selected should cover the longest possible stretch of coastline and should include a wide range of rock types, exposure conditions, and aspects.

(*e*) Destructive sampling should be avoided.

(*f*) Computer storage, retrieval, and processing of the data would be necessary to accommodate the large quantity of information which would be produced.

(*g*) The sampling programme should be planned 12 months in advance. Maintaining regularity would be of prime importance and the investigators would have to give this work absolute priority. In practice this would entail full-time staff without other duties.

Methods

1. General considerations. It was decided to establish a series of sites at suitable places around the coast of Anglesey and in either direction along the mainland coast east and west of the island. These should be easily accessible by road and entail no more than half an hour's walking (generally less). Figure 1 shows the sites being visited at the present time, and table 1 identifies and gives details of these sites. As far as possible, rock faces have been selected which have an even slope, apparent freedom from human disturbance, and an uninterrupted sequence from terrestrial angiosperm vegetation to the *Laminaria* zone.

The basic method chosen was the belt transect, which was considered

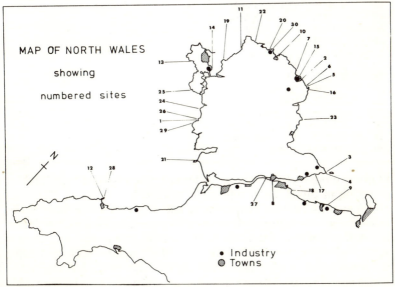

Figure 1 Map of North Wales showing locations of transect sites (see table 1 for details).

Table 1   Details of transect sites shown in figure 1

Site no.	OS grid reference	Name	Siliceous	Calcareous	Very exposed	Moderately ex.	Sheltered	Very sheltered	N	E	S	W
1	SH 330707	Porth Trecastell	+			+					+	
2	453937	Amlwch	+			+			+			
3	641813	Penmon N.		+		+				+		
4	642812	Penmon S.		+			+			+		
5	477931	Porthyrysgaw E.	+				+		+			
6	474933	Porthyrysgaw W.	+		+				+			
7	427945	Bull Bay W.	+			+			+			
8	556714	Menai Bridge	+					+			+	
9	699762	Penmaenmawr	+					+	+			
10	373947	Llanbadrig		+			+		+			
11	299896	Porth Swtan	+			+						+
12	274415	Porth Dinllaen	+			+						+
13	233798	Porth Dafarch	+		+				+			
14	277814	Penrhos	+				+			+		
15	442937	Bull Bay E.	+			+			+			
16	483929	Point Lynas E.	+				+			+		
17	627795	Black Rocks	+					+			+	
18	604730	Penrhyn Castle	+					+	+			
19	283850	Creigiau Cliperau	+			+						+
20	347938	Porth y Pistyll	+				+		+			
21	389633	Llanddwyn	+			+						+
22	317931	Hen Borth	+		+				+			
23	517868	Moelfre		+			+			+		
24	313727	Rhosneigr	+			+						+
25	270748	Rhoscolyn	+			+						+
26	328707	Porth Trecastell W.	+		+							+
27	552716	Church Island	+					+		+		
28	276421	Porth Dinllaen N.	+		+							+
29	329699	Ynysoedd Duon	+		+							+
30	351943	Wylfa	+		+				+			

to give a more useful return in the limited time available than random sampling methods. Each transect is essentially a series of 50 cm by 50 cm quadrat squares placed end to end down the shore. Besides the transects, some quadrats are recorded in greater detail at each visit and a search of a wider area is conducted to note organisms poorly represented on the transects. The choice of 50 cm squares was made after comparative tests of various sizes and is, naturally, a compromise which accepts that a larger, one metre square, unit would be preferable for long algae such as *Ascophyllum nodosum* (L.) Le Jol.

Because of the difficulty in writing under very wet conditions and the need for rapid working, portable tape recorders are used for all data recorded in the field.

2. Setting up the sites. At each site a metric tape was run down the selected rock face and marked by means of a star-chisel at the ends and at the points, at one metre vertical intervals, where the 'permanent quadrats' for detailed recording were to be located. In addition, smaller (20 cm × 30 cm) areas were marked for regular photographic recording in the supralittoral lichen zone.

3. Frequency of visits. Of the 28 sites originally selected, 12 were intended for monthly visits, 7 for three-monthly, 1 for half-yearly, and 8 for annual inspection. Later the number of sites was increased to 30 and the frequency of visits reduced in the case of the three-monthly sites where the additional effort was not found to be justified.

4. Recording methods. These were varied to suit the organisms being recorded. Plants and some sedentary animals were recorded as percentage cover of each quadrat. The method adopted utilized a modification of a pin frame in which optical cross wires were used to provide the recording points (Greig-Smith, 1964). Two point-frequencies were used, 25 points per 50 cm square in the general transect, and 100 points per 50 cm square in the permanent quadrats. The lower intensity allowed the completion of the transect in the time available; the 100-point frame gave greater accuracy, particularly in the case of the smaller organisms. Using different workers in a test, the 25-point frame gave an average error of 4 per cent. In the permanent quadrats, length measurement of some species was included and notes made in all transect quadrats included fruiting condition, hosts and epiphytes, visible damage, bleaching, and presence of young plants.

Animals presented problems which could only partially be met by the use of the belt transect. So far as possible counts of the animals present in each quadrat were made but, for some species, abundance ratings based on the work of previous investigators (Crisp and Southward, 1958; Moyse and Nelson-Smith, 1963; Crapp, 1973) have proved useful.

To overcome the limitations of the belt transect in relation to mobile animals and rarities a search of the shore is also conducted during all visits, normally limited to 15 minutes in a defined area. Data are collected against a check list of about 250 animal species and 160 algae, as well as size classes of some organisms (e.g. *Patella* spp., *Littorina obtusata* (L.), *Nucella lapillus* (L.)), eggs, recently settled spat of different species, and so on.

5. Photography. Photographic recording of the permanent quadrats is undertaken on black and white stock at each visit; this provides a useful safeguard and illustrative material; it is an integral part of the lichen recording method.

6. Lichens. These plants are particularly valuable as indicators of long-term chronic pollution, climatic effects, etc. (Fletcher, 1973a; 1973b) but, being small and slow growing, they present peculiar problems in a surveillance scheme. Most littoral and supralittoral lichen species are encrusting and are therefore very suitable for photographic recording. This is done in 20 cm × 30 cm fixed quadrats, outlined for the purpose of photography by a rectangle scaled in mm so that accurate enlargement can be ensured for subsequent printing and evaluation.

7. Other information. Meteorological data have been collected at each site at each visit and include: air temperature, sea temperature, temperature in upshore pools, sea turbidity, relative humidity, sea state, cloud cover, and wind speed and direction. Besides these, hourly recorded data from Valley RAF station are obtained from the Meteorological Office.

8. Data storage and processing. All collected data are transcribed from the tapes in a standard tabulated form for computer storage, using a standardized numerical equivalent, whatever the recording method used. Each species or other category has a unique number and the data are entered as a matrix of quantity against quadrat number for each species on each site. Data can be retrieved (*a*) in the original form, (*b*) after analysis by simple statistics, (*c*) after analysis by multivariate techniques. For the clearer demonstration of trends one of the most convenient techniques is a contouring programme by which contour lines are drawn through tabulated data at any desired contour interval.

Results

It is proposed here merely to indicate some of the trends which are appearing rather than to present final conclusions; more detailed accounts of the population changes in particular species and communities will appear elsewhere.

1. Seasonal changes. Figure 2 is a contour diagram of changes in percentage cover of *Palmaria palmata* (L.) Kuntze at Porthyrysgaw (Site 5 in figure 1). The species is present in the lower littoral throughout the year, the quantity increasing each summer, though less markedly

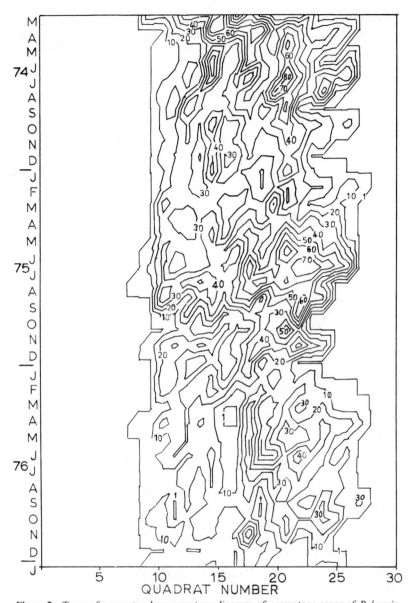

Figure 2 Trace of computer-drawn contour diagram of percentage cover of *Palmaria palmata* along a belt transect consisting of contiguous 50 cm square quadrats at Porthyrysgaw from March 1974 to January 1977. Contours have their values printed alongside them in the original but these figures are too small to reproduce in this reduced copy. The minimum contour is on the left of the diagram; the contour intervals (percentage cover) are 1, 10, 20, 30, 40, 50, 60, 70, 80.

in 1976 than in previous years. Note that the downshore border of the diagram fluctuates—this is merely the result of tidal limitation of the extent to which the shore could be reached on some days. Figure 3 illustrates the same data: here the total *Palmaria* in a permanent quadrat has been averaged by the computer; the major trend is clearly shown but much detail has been lost.

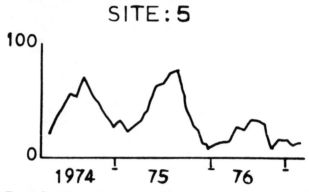

Figure 3 Trace of computer-drawn graph of average percentage cover of *Palmaria palmata* at monthly intervals in a permanent quadrat on the same transect at Porthyrysgaw as in figure 1.

Figure 4 shows the seasonal fluctuation in the ectocarpoid epiphyte *Elachista fucicola* (Vell.) Aresch. Growth begins in spring and maximum density is attained in mid-shore regions in July. There is a tendency for density to increase from the lower towards the higher limits of its range as the season progresses and the population gradually dies away in autumn. Residual plants survived in the mid-shore region into January and February in the first two seasons recorded. In 1976 the quantities were much smaller though the general trend was similar.

In littoral animals the seasonal fluctuations frequently reflect the success or otherwise of settlement from the planktonic larval stage and subsequent survival. Figure 5 shows the fluctuations in *Balanus balanoides* numbers at Porthyrysgaw; the numbers fall towards the end of summer, being replaced by newly settled cyprids in spring. In 1976 recruitment fell markedly, with correspondingly few adults in each class as the season advanced.

2. Catastrophic changes. We were fortunate in late 1974 to have made observations before a man-made change on the shore; subsequently we were able to follow regeneration. In two incidents in November and December 1974 a length of approximately seven metres of the transect at site 2 (see figure 1) at Amlwch was swept clear of most living organisms by the accidental grounding of a steel barge which was being used

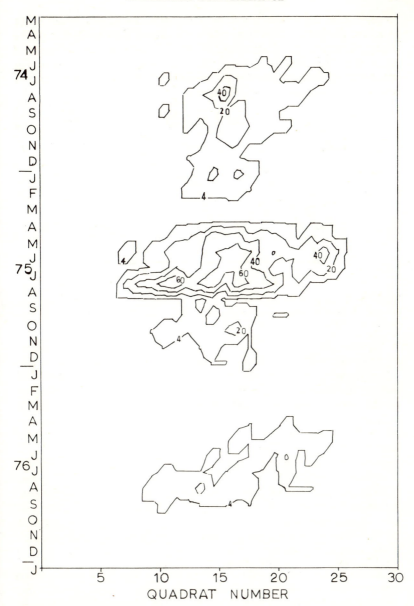

Figure 4 Trace of computer-drawn contour diagram of percentage cover of *Elachista fucicola* along the transect at Porthyrysgaw. For further explanation see figure 1. Contour intervals, from the outermost inwards, are 4, 20, 40, 60.

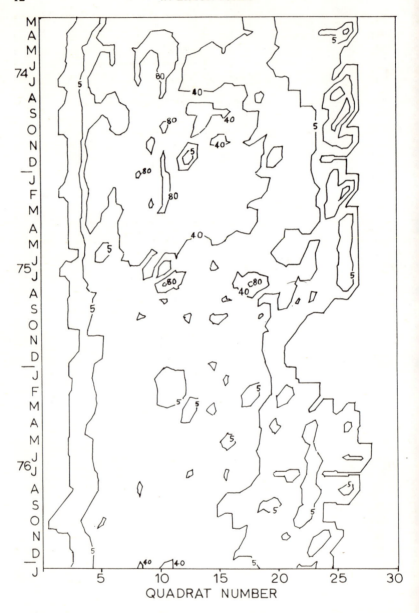

Figure 5 Trace of computer-drawn contour diagram showing changes in adult *Balanus balanoides* along a transect at Porthyrysgaw. For further explanation see figure 1. Contour intervals (percentage cover) as indicated.

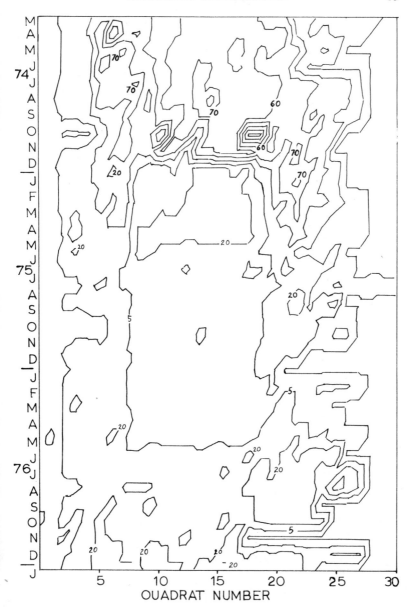

Figure 6 Trace of computer-drawn contour diagram showing changes in the population of *Balanus balanoides* along a transect at Amlwch which was partially cleared by the grounding of barges between quadrats 6 and 20 in November and December 1975. Contour intervals (percentage cover) as indicated.

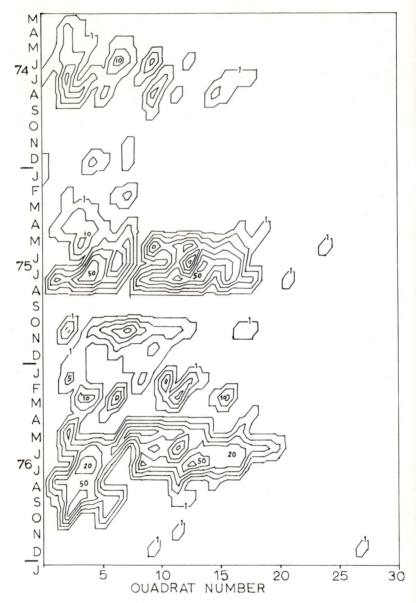

Figure 7 Trace of computer-drawn contour diagram showing changes in the population of *Hyale nilsonii* along a transect at Amlwch partially cleared by barges grounding in November and December 1975. Contour interval (numbers per 0.25 m² quadrat): 0–2, 3–4, 5–9, 10–19, 20–49, 50–200.

by contractors in the construction of new harbour works. Records were subsequently obtained which showed the sequence of regeneration of the populations on the cleared face of rock and these could be compared with the records obtained from the same face in the previous spring. Figure 6 shows the sudden fall in numbers of a sedentary animal species, *Balanus balanoides* (L.), and the slow (and so far, incomplete) recolonization. Filamentous green algae, mainly *Enteromorpha intestinalis* (L.) Link and, later *Cladophora rupestris* (L.) Kütz were the earliest macroscopic colonizers, as demonstrated in figure 7 which shows the increase in the numbers of *Hyale nilsonii* (Rathke), an amphipod associated with such algae (Moore, 1976). The lack of barnacles has been reflected in an absence of *Nucella lapillus* from the transect (figure 8) until new barnacle settlement provided food; an average of two *Nucella* per quadrat were present in December 1976. Limpet numbers, though still lower in 1976 (figure 9) have shown some recovery. Table 2 indicates some changes in algal populations by March

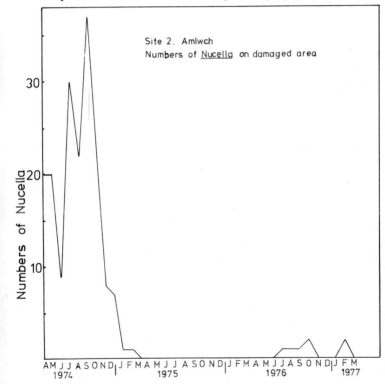

Figure 8 Total numbers of *Nucella lapillus* on the portion of the transect at Amlwch damaged by barges grounding in November and December 1975.

1976; the early colonization by filamentous algae was being overtaken by many young fucoids though there was no sign of the *Ascophyllum* which was originally dominant. Barnacle numbers reached up to 200 per quadrat for both *B. balanoides* and *Elminius modestus* Darwin (which is spreading on to the north coast of Anglesey)—still far below the barnacle cover of 20–30 per cent originally present. One other point worthy of mention is the fate of *Ascophyllum* in the region above the upper limit of damage caused by the barge grounding. *Ascophyllum* persisted at the same density here until April 1976 when it rapidly declined and virtually disappeared. Was this the result of the hot summer of 1976 (if so, it began early at Amlwch!) or perhaps the lack of shelter previously provided by plants lower down the shore?

Table 2   Percentage cover of algae in a permanent quadrat at Amlwch before and after the clearance by grounded barges in November and December 1974

	March 1974	March 1975	March 1976
Ascophyllum nodosum	62	—	—
Lithothamnion agg.	9	—	—
Hildenbrandia spp.	9	1	1
Fucus vesiculosus	40	—	85
Filamentous Green Algae	—	26	—
Porphyra umbilicalis	—	—	58
Filamentous Brown Algae	—	—	13
Cladophora spp.	—	—	8

3. Growth and recruitment. Growth has been recorded in the 'permanent quadrats' either by measurement of the organism (e.g. length in the case of fucoid algae, counting size classes in *Patella*) or by photographic means. At site 5 (the eastern transect at Porthyrysgaw) large numbers of small *Fucus serratus* (L.) plants were present in 1974 (figure 10) and subsequently grew at about 10–15 cm per annum, becoming fewer at the same time. By mid-1976 the comparatively small number of long plants had not been joined by a new generation of small specimens (as might have been expected), recruitment having failed in 1975 on this rather exposed shore. *Fucus serratus* recruitment on more sheltered shores had been more successful in the same period.

On the same shore, the larger size classes of *Patella* declined in number, again without evidence of significant recruitment.

In the lichens, photographic recording has produced ample reliable data but evaluation tends to be slow and costly in labour. However, figure 11 indicates the kind of results which can be obtained by measurement of the thallus in negatives projected to a standard magnification. The encrusting black *Verrucaria amphibia* (R. Clem) is very slow growing; *Parmelia saxatilis* (L.) Ach., a foliose thallus, grows about twice

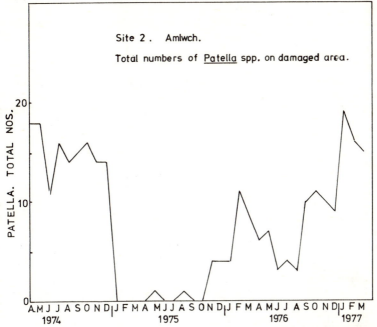

Figure 9 Total number of *Patella* spp. (mainly *P. vulgata*) on the portion of the transect at Amlwch damaged by barges grounding in November and December 1975.

as fast as the others. Seasonal growth is particularly striking in *Anaptychia fusca* (Huds.) Vain where growth is fastest in winter.

4. Community analysis. The stored data can conveniently be analysed various multivariate techniques, the primary intention being to observe possible changes in community structure with the passage of time. Of the methods tried, Principal Component Analysis and Indicator Species Analysis (Hill, Bunce, and Shaw, 1975) proved less useful, because of numerous zero values in our matrices, than the type of Group Average Clustering suggested by Field and McFarlane (1968) and Reciprocal Averaging (Hill, 1973). Group average clustering produces a dendrogram based on the similarity of 'stands' (successive quadrats) and the results of the analysis on two different dates at site 5 (Porthyrysgaw E.) and site 29 (Ynysoedd Duon) are shown in figure 12. It is interesting to note that an objective sampling method has produced (in the March diagram) a clear division of the shore into three zones. The uppermost corresponds closely to the littoral fringe of Lewis (1964) and the community structures described by Russell (1972, 1973, 1977) and leads one to expect that the whole picture corresponds to the

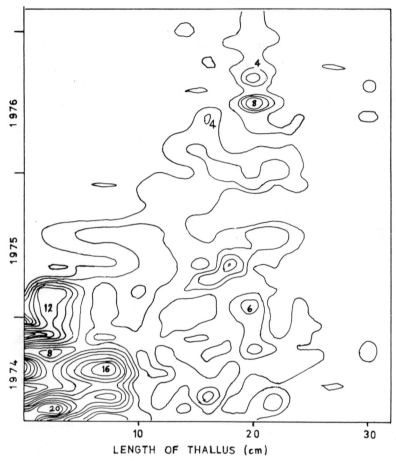

Figure 10 Contour diagram of number of *Fucus serratus* in a series of size classes in a permanent quadrat at Porthyrysgaw. Contour interval: even numbers from 2 to 20. Growth of the surviving plants into 1975 and 1976 was not accompanied by recruitment in these years.

classical picture (Stephenson and Stephenson, 1972) of the three-zone shore. However, the middle group at Porthyrysgaw in March (figure 12a) does not correspond entirely to Lewis's eulittoral for even though *Laminaria digitata* (Huds.) Lamour (the usual marker for the lower edge of the eulittoral) has occasionally been recorded as high as quadrat 20, the bulk of the population of this species begins lower downshore, at about quadrat 28–30 and both *Balanus balanoides* and *Patella* spp. (essentially eulittoral organisms) extend downwards to quadrat 28. In fact, the division at quadrat 21 corresponds rather more closely to the

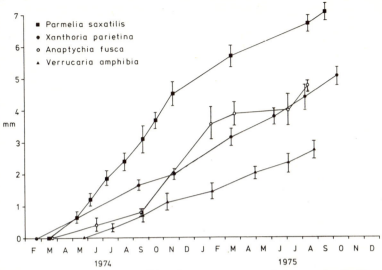

Figure 11 Growth of lichen thalli based on measurements of photographs of a 20 cm × 30 cm quadrat.

upper edge of Lewis's red algal belt; the bulk of the populations of both *Laurencia pinnatifida* (Huds.) Lamour and *Gigartina stellata* (Stockh.) Batt extend from quadrat 20 downwards with only scattered plants above this. Quadrat 20 is also the lower limit for several littoral spp. such as *Nucella lapillus* (in bulk), *Anurida maritima*, and *Enteromorpha intestinalis*. Thus there appears in figure 12*a* to be a clear division of the eulittoral into upper and lower subzones. However, when the results for June are examined (figure 12*b*), it appears that, while the littoral fringe is still clearly defined at the top, the eulittoral comprises three subzones rather than two; the lower division at quadrat 19 is still evident but there is now an upper subzone extending down to quadrat 10. This development is the result of the growth of various summer annuals of limited vertical range and does not imply a fundamental change.

In the two dendrograms obtained from Ynysoedd Duon data in February (figure 12*c*) and September (figure 12*d*) there is a somewhat similar division of the eulittoral into two; in this case the dendrograms are essentially similar, except that in September the better tide allowed quadrats 16, 17, and 18 to be added to the transect. By this time the shore would be assuming its autumnal aspect and most of the summer ephemeral species would be dying down.

Reciprocal averaging has the particular advantage in dealing with data of the kind collected on his project of giving adequate weighting

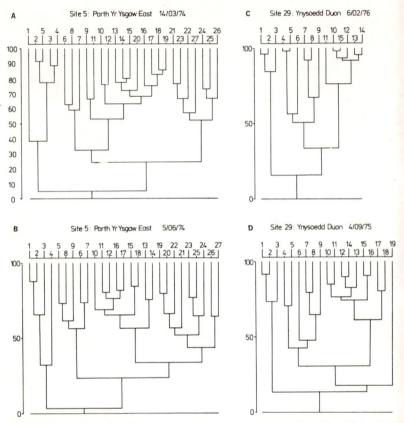

Figure 12 Dendrograms drawn from results of Group Average Clustering of quadrats on the transects on two shores at different seasons. See text for details.

to the distribution of organisms which are present in small quantities. As a result of successive refinements of ordination, the data (in this case quadrats) are ordinated along three axes and can be displayed as a three-dimensional graph. Each quadrat will have a position in this space depending on its ordination along the axes and similarity demonstrated by close proximity within the graph. The method gives better separation of the littoral sub-zones than the group average clustering methods.

5. Trampling. Care has been exercised to avoid damage to the transects; the practice has been to work from the side, avoiding direct contact with the transect. However, this activity has, on a few steeply sloping sites (5 and 15), caused a depletion of littoral life in a band running to seawards alongside the transect. On more gently sloping shores, for example sites 3 and 11, no damage was observed. Where damage occurred, it was visible after no more than a year of monthly visits by two people, suggesting that steep rocky shores are much more vulnerable than we thought. Future danger has been minimized by reducing the frequency of visits in some cases and by extra care by the investigators at the others. Damage usually took the form of a sliding foot scraping off barnacles and other sedentary organisms. The danger to accurate recording of natural changes on the transect is that the strip which has been bared alongside will become colonized by organisms such as *Elminius* or *Enteromorpha* which will spread laterally onto the transect.

Conclusions

Some deductions from the collected and partially processed data have been mentioned above; they are somewhat tentative at this stage and it is anticipated that considerably more will emerge when data for 1977 and 1978 are added, together with detailed information on weather from the Meteorological Office. In terms of the flow diagram proposed by Lewis (1978) we may be said to be involved simultaneously in steps 3, 4, and 5!

However, some points are worth making at the present time:

1. Seasonal effects. Seasonal changes have been noted on all shores, both in species growth and distribution and in community structure. The seasons appear to affect almost all species on the shore and the effects of seasonal change vary greatly from year to year. The exceptionally dry summer of 1976 had a marked effect on littoral life; in general its adverse effect seems to have been more marked in the lower littoral where losses seem to have been greater, proportionally, than in the more desiccation-resistant upshore species. More can be said of this when the performance of various species through the wetter summer of 1977 has been analysed in conjunction with detailed meteorological records.

2. Trampling. The apparent vulnerability of steep rocky shores to damage by sliding boots, particularly in wet weather, indicates that future monitoring programmes must allow for this on any shores where visits are to be more frequent than quarterly. It is also a matter to bear in mind when considering the action to be taken in clearing up after pollution incidents when the increased human activity may lead to serious damage by trampling.

3. Range of variation. While some cyclic changes are already suspected it is advisable to defer comment on this. However, the range of seasonal change from one year to the next has already been found to be large (e.g. 1976 in comparison with 1975) and to raise further doubts on the value of short term 'baseline' surveys.

It has also been found that the range of local variation even on shores separated by short distances makes the extrapolation of results from one to define those on the other a very risky procedure.

4. Analysis of community changes. In view of the apparent variation in community structure resulting from temporary elements in the flora and fauna, it is necessary to exercise caution in interpreting the results of multivariate analysis. It should perhaps be emphasized that the structure of a community on the shore is by no means entirely explicable in terms of floristic or faunistic pattern—when this has been established the underlying physiological causes have still to be determined experimentally. The demonstration of the pattern is merely the beginning of the quest.

Acknowledgements
This paper appears jointly in the names of the present members of the University College of North Wales Coastal Surveillance Unit. We should also like to acknowledge the contribution of members of the unit who have served at various times since 1974; in particular Keith and Susan Hiscock who pioneered the collecting of animal and plant data respectively; Ian Thorburn and Julie Warn who succeeded them; Roderick Shand who began our computer programmes; and also the valuable and willing help of our vacation student assistants.

We are most grateful for the unfailing help of the technical staff of the Marine Science Laboratories and for the continuing support from Shell UK Ltd.

References
Ballantine, W. J. (1961) A biologically defined exposure scale for the comparative description of rocky shores. *Field Studies*, **1** (3), 1–9.

Connell, J. H. (1972) Community interactions on marine rocky shores. *Annual Review of Ecological Systems*, **3**, 169–192.

Crapp, G. B. (1973) The distribution and abundance of animals and plants on the rocky shores of Bantry Bay. *Irish Fisheries Investigations, series B*, **9**, 1–35.

Crisp, D. J. (1964) The effects of the severe winter of 1962–63 on marine life in Britain. *Journal of Animal Ecology*, **33**, 165–210.

Crisp, D. J. and Southward, A. J. (1958) The distribution of intertidal organisms along the coast of the English Channel. *Journal of the Marine Biological Association of the United Kingdom*, **37**, 157–208.

Evans, R. G. (1947) The intertidal ecology of Cardigan Bay. *Journal of Ecology*, **34**, 273–309.

Field, J. G. and McFarlane, G. (1968) Numerical methods in marine ecology. 1. A quantitative 'similarity' analysis of rocky shore samples in False Bay, South Africa. *Zool. africana*, **3**, 119–137.

Fletcher, A. (1973a) The ecology of marine (littoral) lichens on some rocky shores of Anglesey. *Lichenologist*, **5**, 368–400.

Fletcher, A. (1973b) The ecology of maritime (supralittoral) lichens on some rocky shores of Anglesey. *Lichenologist*, 5, 401–422.

Grieg-Smith, P. (1964) *Quantitative Plant Ecology*, 2nd edn. London: Butterworth.

Hill, M. O. (1973) Reciprocal averaging: an eigenvector method of ordination. *Journal of Ecology*, 61, 237–249.

Hill, M. O., Bunce, R. G. H., and Shaw, M. W. (1975) Indicator species analysis. *Journal of Ecology*, 63, 597–614.

Jones, W. E. (1974) Changes in the seaweed flora of the British Isles. In *The Changing Flora and Fauna of Britain*, ed. Hawksworth, D. L. Ch. 7. pp. 97–113. London: Academic Press.

Jones, W. E. (1978) A survey of the marine biology of Bardsey. *Bardsey Observatory Report*, 21 (in press).

Lamb, M. H. (1972) *Climate Past, Present and Future*. London: Methuen.

Lewis, J. R. (1964) *The Ecology of Rocky Shores*. London: English Universities Press.

Lewis, J. R. (1976a) The role of physical and biological factors in the distribution and stability of rocky shore communities. In *Biology of Benthic Organisms, Second European Symposium on Marine Biology*, ed. Keegan, B. F., Ceidigh, P. O., and Boaden, P. J. S. pp. 417–424. Oxford: Pergamon.

Lewis, J. R. (1976b) Long term ecological surveillance: practical realities in the rocky littoral. *Oceanography and Marine Biology Annual Review*, 14, 371–390.

Lewis, J. R. (1978) The implications of community structure for benthic monitoring studies. *Marine Pollution Bulletin*, 9, 64–67.

Lewis, J. R. and Bowman, R. S. (1975) Local habitat-induced variations in the population dynamics of *Patella vulgata* (L.). *Journal of Experimental Marine Biology and Ecology*, 17, 165–203.

Moore, P. G. (1976) Organisms in simple communities: Observations on the natural history of *Hyale nilsonii* (Amphipoda) in high littoral seaweeds. In *Biology of Benthic Organisms, Second European Symposium on Marine Biology*, ed. Keegan, B. F., Ceidigh, P. O., and Boaden, P. J. S. pp. 443–451. Oxford: Pergamon.

Moyse, J. and Nelson-Smith, A. (1963) Zonation of animals and plants on rocky shores around Dale, Pembrokeshire. *Field Studies*, 1 (5), 1–31.

Russell, G. (1972) Phytosociological studies on a two-zone shore, I Basic pattern. *Journal of Ecology*, 60, 539–545.

Russell, G. (1973) Phytosociological studies on a two-zone shore, II Community structure. *Journal of Ecology*, 61, 525–536.

Russell, G. (1977) Vegetation on some rocky shores at some North Irish Sea sites. *Journal of Ecology*, 65, 485–495.

Stephenson, T. A. and Stephenson, A. (1972) *Life Between Tidemarks on Rocky Shores*. San Francisco: W. H. Freeman.

Monitoring the effect of oil pollution on rocky seashores

A. NELSON-SMITH

Department of Zoology, University College of Swansea, Swansea, Wales

Some basic difficulties

Any pronouncement about the effects of pollution upon the sedentary life of a stretch of seashore implies a knowledge both of the previous, untainted state of that shore and of the magnitude of changes which would have taken place during the period of pollution, but from purely natural causes. It is often impossible to demonstrate that significant long-term changes have taken place for lack of suitable base-line data of this sort. Many forms of marine pollution have exerted their influence at a low intensity for such a period of time that the shore biota may not have been observed in their previous unaffected state or, perhaps at best, such observations may have been recorded only in a manner which does not permit careful comparison with their present condition. Considerable lengths of coastline may never have been studied at all and, although a competent shore ecologist might be able to make reasonable assumptions from his knowledge of nearby areas in such regions as north-western Europe, the Mediterranean, or North America, this is unlikely to be helpful further afield. Even where clearly defined changes have been reliably demonstrated, it may be difficult to assign them to a particular influence (such as a polluting discharge) when others (for example, gradual climatic changes, unusual extremes of temperature, epidemics of disease, or invasions by predators, parasites, and exotic competitors) could have had as powerful an effect. Tides and currents are often so complex that it is also difficult to show exactly which localities might be affected by a discharge whose toxic constituent is usually detectable only by using elaborate instruments in the field, or by taking samples for subsequent analysis in the laboratory.

In concentrating on the rocky intertidal zone around the British Isles, I am taking a relatively soft option in that all but the smallest and most specialized of its plants and animals are spread out for inspection and enumeration without the necessity of separating them from a habitat which normally conceals them (as in soft sediments), or suffering undue restrictions on the method and extent of sampling (as with the sub-littoral benthos, plankton, or the larger organisms swimming in open waters). Such shores are probably more completely described than any others, yet there are still vast areas of ignorance of even their most basic ecology. In the same way, crude oil and most of its products provide the investigator with an advantage over most other pollutants in that (except in the vicinity of refineries and other installations discharging a waste stream which may be only lightly contaminated) they are usually immediately obvious and often cover the shore in quantities which will plainly have more effect than any other factor except, perhaps, the measures used to clear them up. Damaging spills of fresh oil occur most frequently in oil ports where, even though facilities to contain, recover, or disperse it may be readily available, considerable amounts will regularly be stranded in locations which can to some extent be predicted. However, the location of the most dramatic accidents has depended on unpredictable errors in navigation, mechanical failures, or extreme weather and sea conditions, so that the spilt oil affected shores remote from the port to which it was consigned or from which it was shipped. *Torrey Canyon*, bound for Milford Haven from the Middle East, oiled beaches in Cornwall and Brittany in 1967; *Metula* spilt her cargo in the Straits of Magellan whilst bound for the Pacific coast of Chile during 1974; *Showa Maru* was returning to Japan with a load of crude oil in 1975 when she became wrecked in the Strait of Singapore; and *Amoco Cadiz* was carrying crude oil to Rotterdam but severely polluted the Brittany coast early in 1978. Numerous others have provided less well-publicized examples during the last decade. Thus it may be that a shore becomes heavily polluted without having been adequately studied in the past, and without there being sufficient warning to permit the organization of even a hasty survey before the onset of pollution. Oil from *Torrey Canyon* reached none of the shores surveyed in order to establish its effect on them although some of the shores which it did damage have been studied extensively for other purposes (see Smith, 1968). I shall therefore consider first of all the uses which can be made of work which was originally carried out for purposes other than the monitoring of pollution before discussing the applicability of various methods specifically for this purpose.

Fauna and flora listing

The simplest way of reporting on the biology of a shore is to provide a fauna and/or flora list. Many of these are available, from Victorian times onwards; they may consist of an exhaustive list for a particular locality or be restricted to a particular group of plants or animals, often studied over a wider area. In the latter case, some localities may receive very cursory treatment: for example, in a report on fauna of the northern shores of the Bristol Channel (Purchon, 1957) only three mollusc species are listed for Swansea Bay—an expanse of sheltered and rich muddy sand from whose strandlines alone I have collected the empty shells of twenty or thirty different species within half-an-hour. The remaining invertebrates included in the list comprise one sponge, one polychaete, two crabs, and a brittle-star. The taxonomic status of many entries will have changed either by the 'lumping' of Victorian species distinguished only by superficial characters such as colour-pattern, or the 'splitting' of old species now recognized to contain several different physiological types. Conclusions based on an increase or decrease in the number of species since a previous list was made can then be drawn only if one is expert in the taxonomy of the groups included. Such lists naturally reflect the particular interest of their compilers, being virtually complete for one group but scanty for others. Comparisons may in these circumstances be very deceptive: it might have been concluded from one aspect of the investigation of possible effects of a serious oil-well blow-out in the Santa Barbara Channel, California (Nicholson and Cimberg, 1971) that the smaller red algae had actually benefited from heavy shore pollution, had it not also been reported that the base-line surveys were carried out, over the years, by unspecialized biology students, whereas the post-spill survey had been made by a researcher particularly interested in the Rhodophyceae.

Where faunistic records are maintained over a considerable length of time, it is natural that the numbers of species listed should continually increase; this did not inhibit one worker (Arnold, 1959) from concluding that oil pollution from shipping in Plymouth Sound was not having serious effects because successive editions of the *Plymouth Marine Fauna* (last published by the Marine Biological Association in 1957) each included a greater variety of organisms. Indeed, one way of determining the diversity of life in a given habitat is to record the number of different species discovered in successive short periods (say, half hours) of search. If the limitations of such comparisons are borne firmly in mind, they may still be of use: for example, Rattray's list of the algae of the Firth of Forth (1886) includes several important species which are certainly not still present, indicating that conditions there have in some way become less favourable. Over a shorter time-span, Dias's list of plankton netted from various parts of Milford Haven just prior to its rapid growth into a major oil-port compares well with catches made by Gabriel ten years later (Gabriel, Dias, and Nelson-Smith, 1975), suggesting that the impact of oil-spills on this community of organisms has not been severe. However, it is recommended that, in general, such listing before and after any development which might be expected to have ecological effects should, for preference, be compiled on each occasion by the same worker and (unlike most early lists) should comment wherever possible on the precise location and zone occupied, the abundance or density of cover, the size, maturity or reproductive state, and apparent health of the organisms included.

Study of settlement surfaces

Special panels

For many years, panels or blocks of timber, asbestos-cement, slate, earthenware, or various plastics have been suspended from rafts or fixed to solid surfaces to collect the settling stages of algae and sedentary animals as an indication of the severity of biological fouling to which ships' hulls, buoys, harbour structures, and sea-water conduits might be subjected at that location, or to determine the possible intensity of attack by wood-boring organisms. Typical installations and results from all around the world are illustrated in the manual prepared for the US Naval Institute by Woods Hole Oceanographic Institute (1952) or (in the case of wood-borers) in an OECD handbook (Jones and Eltringham, 1971). Settlement panels have also been exposed in one or two instances to assess the potential of commercial shellfisheries and Hillman (1975) has suggested their use for more general environmental monitoring.

Rocky shores are very variable in their exposure to wind and wave action, aspect, and hence degree of insolation or dampness, slope, and surface texture. Conditions at the top and bottom of the intertidal zone differ widely enough to give rise to the familiar phenomenon of a marked vertical zonation of the sedentary biota. Panels of selected texture, on the other hand, can not only be placed at any desired orientation and tidal level (or become independent of tidal emersion by attachment beneath a buoy or pontoon), but can also be set out at any season and left on site for any chosen interval of time. After recovery, a panel may be removed to the laboratory with minimal disturbance to the attached biota, all features of which can then be investigated at leisure. It is possible to allow a panel to acquire biota typical of one (say, unpolluted) locality, or even to settle it exclusively with individuals of a selected species in the laboratory, and then place it in a site thought to be polluted or with some other special characteristics in order to determine the response of such species not already present there. This flexibility is very attractive for the study of fixed discharges or more general experimental purposes but, of course, the method can be used to investigate the effects of an accidental oil-spill only if panels have been placed in the right locations at some time in advance of the accident. Numerous panels may be lost as a result of bad weather, other accidental damage, or public interference ; a full programme of seasonal sampling, allowing for proper replication of results as well as such losses, calls for many panels and a correspondingly large amount of work, first in making up and fixing the frames needed to support them and then in removing, scrutinizing, renovating, and replacing the individual panels appropriately. In an investigation of the possible effects of heated power-station effluent in Long Island Sound described by Hillman (1975), a total of 84 panels was deployed at any one time on six sites over a five-year period. The system of sampling called for the inspection of nearly 800 panels during that time; the number of replacements made necessary by loss and damage was not given, although such problems were mentioned.

Settlement panels cannot, moreover, become a complete microcosm of shore life. As they are usually attached at some distance from the actual rock surface they will, for example, receive only those spores or larvae which are free-swimming or long-drifting; in the absence of competition, those organisms which do settle may show abnormally high growth-rates. Mobile but non-swimming grazers and predators such as limpets, whelks, and various echinoderms might be unable to reach them (or, if they do so, a single panel is likely to prove too small a territory for the maintenance of a recognizable ecological balance), although fish would often find such panels easier to feed from. The absence of crevices, larger irregularities, and small trapped bodies of water during periods of tidal emersion, together with projection further into the region disturbed by wave-action and currents, may reduce the amount of protection normally available to sedentary forms, although the food supply to filter-feeders could be enhanced. Because of such oversimplification of the physical as well as biotic features of the environment, great care is needed in deriving assumptions about actual shore conditions from the data obtained in settlement-panel studies.

Other artificial surfaces

Artificial structures not primarily intended for these purposes may nevertheless provide useful sampling areas of standardized if not wholly natural characteristics, often offering a compromise between small panels exposed for short periods and the intertidal zone of the shore itself by virtue of their greater size and permanence. For example, a series of navigation buoys in Plymouth Sound and up the River Tamar provided Percival (1929) with a means of investigating the penetration of various marine species into increasingly estuarine conditions in a situation where natural hard surfaces are lacking for considerable distances (also, incidentally, eliminating the necessity of allowing for the diminution of tidal amplitude in interpreting his results). Corlett (1948) sampled jetty-piles in his study of similar penetration into the Mersey Estuary. The legs of oil-rigs offer support for sedentary biota in offshore areas, where they are not only the sole available hard surface but also occur in the immediate vicinity of likely sources of pollution. Unfortunately, rivers and small estuaries are decreasingly used by shipping, so that fewer buoys are available; these are now repainted with quick-drying products on station, depriving the modern investigator of Percival's opportunity to scrutinize them at leisure in the maintenance yard. Of course, it is also the aim of many operators of such structures specifically to discourage the development of fouling growths, thus further decreasing their comparability with a natural habitat. However, stone and concrete quaysides or breakwaters resemble very closely steep rockfaces and, again, are often sited where pollution is common. By referring to observations of such surfaces, made over the years in the harbour at Marseille, Gilet (1959) was able to demonstrate the deleterious effects of chronic oil pollution there. In a much overdue base-line study of rocky-shore biota between Worm's Head and Nash Point, on the Welsh shore of the Bristol Channel, sections of sea-wall and breakwaters had to be included because of the lack of such natural shores at the head of Swansea Bay (Nelson-Smith, 1974). At a more academic level, Lysaght (1941) and Southward and Orton (1954) used the surfaces of the Plymouth breakwater—constructed from natural rock but much more precisely vertical and offering more readily defined exposure to or shelter from oceanic swell than nearby natural shores—in studies of a small gastropod, or most of the common intertidal biota, respectively.

Wider studies of shore biota

Wherever possible, then, the best and most reliable way of assessing the ecological impact of an oil-spill or persistent outfall of oily water on the intertidal zone is to compare data, obtained before and after the onset of pollution. These must be more informative than a mere list of presence or absence and have been gathered from study of the shore itself rather than from any artificial surface. As has already been pointed out, a great variety of published work already forms a basis for this to some extent, dealing with many sections of the coastline of northwestern Europe and, much more patchily, further afield. Such papers make up the backbone of the voluminous literature from which the present generation of seashore biologists learned their craft and, of course, it is quite impossible to mention more than a small selection here. The original intentions of the authors of such studies fall roughly into three groups. Many have concerned themselves with the detailed description of a small stretch of coast; examples are provided by Evans (1947, 1949) and Lewis (1953, 1954). Others have been more interested in the disposition of a particular group, perhaps throughout a wider region—these are typified by Naylor's work (1930) on two maritime lichens around Plymouth; Moore's numerous papers of 1934–1940 on British barnacles and gastropods (to which I give no detailed reference here, but which all appeared in the *Journal of the Marine Biological Association of the UK*); Kitching's summary (1950) of the distribution of the barnacle *Chthamalus*, also around the British Isles, updated by Southward and Crisp (1954); or Crisp (1958) on the spread of the immigrant barnacle *Elminius* throughout north-western Europe. Finally, a number of workers have built upon such studies wide-ranging schemes of intertidal zonation and geographical distribution, culminating in the books of Lewis (1964) on the British coast and of Stephenson on rocky shores throughout the world, published posthumously in 1972. Apart from the contribution originally intended, these provide details of the hundreds of publications which I have been unable to include in this brief review.

The first category of such studies may prove almost ideal for our purpose, should the oil-spill occur or the outfall be sited across one of the shores described in this way. At best, the data are semi-quantitative and distinguish between community structure and abundance at different vertical levels on the shore, so that it may be possible at any time to repeat the work, using the methods described by the original investigator and obtaining a directly comparable before-and-after result. Unfortunately, only a tiny proportion of the coastline at risk has been so described. Some accounts, although valuable for their intended purpose, are idiosyncratic in such a way that a re-survey of the shore could be undertaken only by the original author—for example, the ecology of some Galician rias has been presented (Ardré *et al.*, 1958; Fischer-Piette and Seoane-Camba, 1962, 1963) in an elegant but quantitatively rather uninformative narrative manner, as though guiding a field excursion. Many valuable studies of seashores have never appeared in print—perhaps because, during a recent period in the development of marine biology, laboratory-oriented work was more fashionable and many editors underestimated the usefulness of reports of field surveys, especially of a general nature. Such unpublished data can usually be compared with later results only by those who originally collected them (as D. P. Wilson was able to do during the oiling of Trevone, on the north Cornish coast—see Smith, 1968) and subsequent workers may not even be aware of their existence.

Studies of a particular species may give quantitative data on abundance (or some other useful parameter) at a large number of stations around a coastline, but in a manner unsuited to the purpose under present consideration—for example, the maximum density on each shore, which may be achieved at different levels not equally susceptible to a minor oiling, or an average for the whole intertidal range. The presence of other species, even if important in a more general sense, may go unreported. Again, more complete data might have been recorded but not published and will thus not be generally available. The value of such accounts may depend on the wider applicability of the methods used, or on whether the species dealt with becomes recognized as an 'indicator' for effects of a particular nature (see below).

General schemes of zonation and distribution are, by their nature, of little use in providing details of the unpolluted state of particular localities but they offer an established framework of nomenclature, system of enumeration, or method of working which, if adopted by all investigators in this field, would make it easier both to understand and compare their reports. Such schemes indicate very broadly what might be expected on a given type of shore and, in dealing with the serious pollution of a substantial sector of coastline in one of the less studied parts of the world, may provide the only information available to an investigator other than his own observations.

The belt-transect/abundance scale method

The zonal distribution of rocky-shore biota demands systematic enumeration throughout the whole intertidal zone, which immediately discounts random sampling methods such as the casting of quadrats. The structure and physical characteristics vary across as well as down most shores, although in a less systematic manner, and they are occupied in a patchy distribution by many of their sedentary inhabitants. These may be colonial or solitary, numerous or sparse, tiny and half-hidden or draped across the surface in long fronds from a single point of attachment. The seemingly precise method of a line-transect (in which every organism touched by a line stretched in a chosen direction is recorded but others are ignored) or the use of a points-frame along a line of stations (when those organisms touched by a pointer deployed according to a set pattern are similarly recorded) might thus give quite unrepresentative results. More flexible methods seemed to be called for when, in 1960, I set out to describe the basic ecology of rocky shores in Milford Haven (an extensive deep estuary on the south-western corner of Wales which has since become Britain's largest oil-port), partly from academic interest but mainly so that the effects of operating the oil installations and related activities could be determined in later years. My colleague John Moyse and I drew on some of the earlier work mentioned here in devising a speedy but reasonably repeatable method for surveying belt-transects (in which a broader band of the shore is scanned). We used this first around the Dale peninsula (Moyse and Nelson-Smith, 1963) and I then extended the surveys to the remaining shores around the mouth and up the estuary itself (Nelson-Smith, 1965, 1967).

The basic method

Because time on the shore is inexorably limited by the incoming tide, we restricted the list of species for enumeration to about sixty plants and animals which are common, important, and easy to distinguish *in situ*—although not all of these would be present on a single shore. These were divided into a few groups, each containing organisms whose typical abundance or extent of surface cover is similar; suitable criteria of abundance for each group were used to define a short, graded scale (table 1) so that their density could be assessed rapidly and recorded simply, rather than in absolute numbers per unit area. The transect itself should be visualized as a band about 20 ft (6–8 m) wide extending

Table 1   Criteria of abundance for common plants and animals of rocky seashores. Originally taken from Crisp and Southward (1958), modified and extended by Ballantine (1961), Moyse and Nelson-Smith (1963), and Crapp (1973).

R—rare	O—occasional
F—frequent	C—common
A—abundant	S—superabundant
Ex—extremely abundant	

1. Barnacles (except *Balanus perforatus* and *Littorina neritoides*)
   - Ex   more than 5 per cm^2
   - S   3–5 per cm^2
   - A   1–3 per cm^2/more than 1 per cm^2*
   - C   10–100 per dm^2
   - F   1–10 per dm^2 (but never more than 10 cm apart)
   - O   10–1000 per m^2 (few within 10 cm of each other)
   - R   fewer than 10 per m^2

2. *Balanus perforatus* (and less common sublittoral barnacles)
   - Ex   more than 3 per cm^2
   - S   1–3 per cm^2
   - A   10–100 per dm^2/more than 10 per cm^2*
   - C   1–10 per dm^2
   - F   10–100 per m^2
   - O   1–10 per m^2
   - R   fewer than 1 per m^2

3. Limpets and winkles other than *Littorina neritoides*
   - Ex   more than 200 per m^2
   - S   100–200 per m^2
   - A   50–100 per m^2/more than 50 per sq m^2*
   - C   10–50 per m^2
   - F   1–10 per m^2
   - O   1–10 per dam^2
   - R   fewer than 1 per dam^2

4. Topshells, dog-whelks, anemones (also suitable for sea-urchins, oysters etc.)
   - Ex   more than 100 per m²
   - S   50–100 per m²
   - A   10–50 per m²/more than 10 per m²*
   - C   1–10 per m² generally ⎫ may be more very locally
   - F   less than 1 per m² generally ⎭
   - O   always less than 1 per m²
   - R   fewer than 1 per dam²

5. Mussels
   - Ex   more than 80% cover
   - S   50–80% cover
   - A   20–50% cover/more than 20% cover*
   - C   large patches but less than 20% cover
   - F   scattered in small patches
   - O   scattered individuals but more than 1 per m²
   - R   fewer than 1 per m²

6. Large tubeworms (eg *Pomatoceros* and *Sabellaria*)
   - A   more than 50 tubes per dm²
   - C   1–50 tubes per dm²
   - F   10–100 per m²
   - O   1–10 per m²
   - R   fewer than 1 per m²

7. Small tubeworms (eg *Spirorbis* and *Polydora*)
   - A   more than 5 per cm² generally *or* on 50% or more of suitable algae
   - C   more than 5 per cm² locally *or* less than 50% of suitable algae
   - F   1–5 per cm²
   - O   fewer than 1 per cm², more locally
   - R   fewer than 1 per cm² generally

8. Sponges, *Catenella*, lichens, and flowering plants
   - Ex   more than 80% cover
   - S   50–80% cover
   - A   20–50% cover/more than 20% cover*
   - C   5–20% cover
   - F   large scattered patches but less than 5% cover
   - O   small scattered patches
   - R   less than one small patch or plant per m²

9. Algae other than *Catenella*
   - Ex   more than 90% surface cover
   - S   60–90% surface cover
   - A   30–60% cover/more than 30% cover*
   - C   5–30% cover
   - F   less than 5% cover but zone still apparent
   - O   scattered plants, zone indistinct
   - R   less than one small patch or plant per m²

*When not using S and Ex grades.

from the lowest station, as deep in the water at low tide as the investigator can work efficiently (perhaps 15–20 cm), up to the first point that can be regarded as truly terrestrial (usually taken as the level of the first few flowering plants). Such a transect is divided into stations at equal vertical intervals because, of course, zonation is essentially a tidal phenomenon and is thus a function of water level rather than distance measured horizontally or down the slope of the shore. In these surveys, stations were marked at two-foot vertical intervals largely as an arbitrary matter of convenience, at first using an instrument made from slotted metal angle which we termed a 'gallows', in which the foot of a limb 2 ft (61 cm) long was placed at the station under investigation while a longer limb, bolted at 90° to its upper end and containing a bubble-level, was used as a sighting-bar to seek a suitable spot for the next station. The line of stations, which need not necessarily run straight or remain exactly at right-angles to the water-line, is taken as the midline of the transect.

Surveying usually starts from the bottom of the shore and the absolute level of each station is assigned on the assumption that low water is at the level predicted in reliable tide-tables. At each station, the rock surface is examined across the full width of the transect within horizontal limits which are judged to fall half-way vertically between the marked spot and the stations next above and below it. The abundance or density of cover recorded for each listed organism is that which is deemed to be typical of, or averaged over, the whole area (according to whether the organism in question is discontinuously grouped or wide-spread across it). When making academic ecological studies, the occupants of deep crevices, rock-pools, the underside of boulders, or the shoreward surface of stacks and pinnacles are ignored or recorded simply as 'present', as these regions are not typical of the shore in general; exceptions must be made for a few species, such as the tiny snail *Littorina neritoides* whose preferred habitat is the pits and crevices which have often been excavated by past generations of the same species. Zones may occasionally be so narrow (as with the band of the brown alga *Pelvetia canaliculata* which usually grows at the level of neap-tide high water) or transitions so rapid (such as the 'barnacle line' at much the same level, above which markedly fewer of the acorn-barnacles *Balanus balanoides* are to be found) that assessments for intermediate levels may occasionally have to be recorded.

It is useful, particularly in order to help in explaining the sudden changes in community structure which occasionally become clear only when plotting the results graphically after leaving the transect, to note at each station the slope, texture, and nature of the surface, together with special features such as the proximity of potentially abrasive sand or shingle patches, regions affected by salt- or fresh-water drainage, and those experiencing greater exposure or shelter than the shore as a whole. Location of the transect for repeat surveys is facilitated by making a permanent mark at the topmost station, which should be beyond the reach of most wave-swash; on well-colonized shores, there is no easy way of marking intertidal stations so that they can readily be found again after more than a few weeks. When the surface is moderately dry, an aerosol spray-can of automobile paint provides satisfactory (if, in a minor way, anti-social) temporary markings; amongst plentiful seaweeds, one or two easily visible, heavy, and stable markers (such as light-coloured cobbles from the head of the beach) are useful to remind one where the centre-point is of each station as the survey proceeds. When it is intended to revisit the transect regularly and the line of stations is reasonably straight, a literal 'transect line' can be prepared from strong, light cord knotted in such a way that, when one end is clipped to a ring-bolt sunk into the rock at a suitable top-shore location and is then deployed in the correct direction, each knot lies at one of the original survey stations. An important piece of practical advice is to establish the first few stations on the lower shore as soon as the tide has ebbed to its limit, before the base-mark of the survey becomes too deeply inundated by the flood. It is, of course, possible to reverse the survey process and work down the shore, either taking as datum a previous high-water mark or noting the relationship of the lowest station with the level reached by low water. Careful planning, experience in the method and prior knowledge of that sector of the coastline make it possible to survey a transect up one shore during the flood tide and another down an adjacent shore during the following ebb in the course of a single day.

The 'gallows' requires two operators, one to keep the sighting bar horizontally level while the other looks along it. In practice, surveys are best conducted by two workers for other reasons: for example, one can search amongst seaweeds and boulders while the other makes notes with dry, clean hands. On the lower part of a very exposed shore, the recorder is able to warn the searcher of the approach of waves, while mutual assistance may be necessary in scrambling over the steeper upper parts. Throughout the survey, the enumeration of organisms at each station can be carried out more efficiently and completely if the recorder prompts the searcher from the list given on the survey record-sheet. However, there are many occasions when an investigator has to survey a transect without assistance, so a modified instrument was developed which became known as a 'cross-staff' (see Nelson-Smith, 1970, and figure 1). In this, the horizontal bar is much shortened (consisting originally only of a medium-sized wooden bricklayer's level) and is mounted symmetrically on the vertical member; this is extended above the bar to accommodate and protect an angled mirror

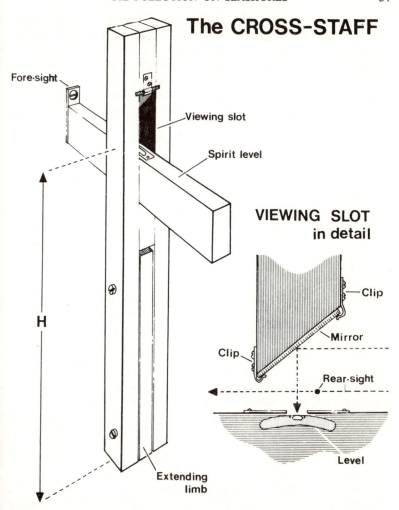

The CROSS-STAFF

Fore-sight

Viewing slot

Spirit level

VIEWING SLOT
in detail

Clip

Mirror

Clip

Rear-sight

Level

H

Extending
limb

Figure 1   The 'cross-staff' used in surveying belt-transects. H (the station interval) is best adjusted to about one-tenth of the mean spring-tide range. From Nelson-Smith (1970).

which enables the user to watch the bubble-level as he sights along the bar. To compensate for the greater possibility of slight angular error introduced by using a short cross-bar, a pair of simplified sights is fitted.

Shore surveys are normally made during spring tides, when the lower shore is uncovered to its greatest extent. The range of such tides in Milford Haven and elsewhere in the south-west where surveys of this sort were first made is 20 ft or more, so that about ten intertidal stations were established on each transect when using a 2 ft vertical interval. The experience of subsequent workers when surveying shores with a smaller tidal range—for example in Bantry Bay, Ireland (Crapp, 1973), Shetland (Addy *et al.*, 1973), and Norway (E. B. Cowell, personal communication)—is that this is about the best number. It is therefore recommended that the vertical interval should always be adjusted to one-tenth of the mean spring-tide range and later versions of the 'cross-staff' have been constructed with adjustable or interchangeable lower limbs. During his work in Bantry Bay, Crapp found that limpets and some barnacles are present there in numbers considerably higher than those envisaged when the criteria of abundance were originally adopted. On advice from J. R. Lewis, he therefore introduced first 'super-abundant' (S) and then 'extremely abundant' (Ex) grades not only for these, but also for the remaining groups; they have been included in table 1.

Error and repeatability

From this account of the survey method, it will be clear that several sources of error or disagreement exist. Cartographic benchmarks or other means by which the comparative levels measured up the transect could be checked or converted to absolute heights are only rarely found within a reasonable distance. The tidal predictions used to indicate the height of the first station above Chart Datum (which itself is not at the same level at all points along a lengthy coastline) assume normal baro-metric pressure and still air; even under these conditions, the position of the water's edge may be difficult to judge precisely if it is much disturbed by breaking waves. Fortunately, absolute levels are important mainly when making comparisons between distant shores for academic purposes; when comparing similar, nearby shores or returning later to the same transect, the levels characteristic of certain biota themselves —for example, the *Pelvetia* zone or the 'barnacle line' mentioned above —usually provide adequate calibration. The angular error of the sight-ing procedure increases with horizontal distance between stations; the crude instruments described here would be of little use on a gently-sloping sandy beach but, whenever a transect has been surveyed from low water to the known level of a previous high tide on rocky shores of typical structure, the disparity is rarely more than a few inches. Instru-mental errors, and those introduced in selecting and marking individual stations, are as likely to be positive as negative and thus tend to balance each other out.

Although it must be taken for granted that the investigator is capable of identifying correctly *in situ* the species which he has to enumerate, the method also requires him to make judgments which depend largely on experience—for example, in choosing the best line for a transect and selecting the exact site for each station along it, deciding which areas or organisms to ignore (or record only as 'present') because they are not typical, and assigning abundance grades to those organisms which he considers worthy of inclusion. Where the density of a given plant or animal has a value lying on the boundary between two abundance grades, it is obviously possible to assign it with equal validity to either. An experiment was carried out (Baker, 1976) in which the same Milford Haven transect was surveyed independently, but within a few days, by two graduate members of the staff of the nearby Oil Pollution Research Unit at Orielton who are well experienced in making such surveys and an undergraduate with only a few weeks' experience. Assessments differed between the three, occasionally to the extent of one grade of abundance and, exceptionally, even more; in a few cases, the limits of vertical distribution assigned to individual organisms differed by one station (in this case, 0.5 m). The differences seem to have occurred mostly when dealing with species which are small, partially hidden, or difficult to separate taxonomically—such as *Littorina neritoides*, crevice-dwelling sea-anemones, or acorn barnacles respectively. Overall agreement was surprisingly good. Nevertheless, the most reliable comparisons (whether between sites or of successive surveys at the same site) are obviously to be obtained when the same well-trained and experienced worker has undertaken each survey.

Use in pollution incidents

Baseline surveys in Milford Haven

Whilst it was recognized that any of the transects established around Milford Haven might become oiled, two (at Hazelbeach on the north and Llanreath on the south side) were considered of special interest because they are the nearest rocky shores to the point of discharge of the heated effluent from the Pembroke Power Station, whose construction had not then started. In the hope of being able to distinguish possible effects of the warm water from purely seasonal changes, these transects were re-surveyed in spring and autumn for several years, revealing changes on a scale no greater than between the three participants in the comparability experiment already referred to. However, before the Power Station came into use, both shores had been seriously polluted by crude oil from a damaged tanker and, less than two years later, Hazelbeach was again badly affected by an overflow of similar oil from the tank-farm of a neighbouring refinery. On at least the first of these occasions, damage to the biota was compounded by the use of a highly toxic 'first-generation' dispersant in washing off the oil. A repeat survey of each shore when conditions had initially stabilized, about a month later, revealed changes both in the vertical range occupied by many species and in abundance at each station which were of far greater magnitude than could be the result either of natural fluctuations or of the various sources of error discussed above.

The Oil Pollution Research Unit was set up at Orielton Field Centre in 1967 by the efforts of E. B. Cowell, then its Warden, following the first of these oiling incidents. G. B. Crapp, one of its first staff members, undertook repeat surveys of my transect sites in 1968–1970, after which occasional (but so far unpublished) later surveys have been carried out by more recent members of the Unit's staff. Hazelbeach is a moderately sheltered shore of stable boulders and smaller rocks as well as bedrock ridges, supporting large populations of gastropods. A comparison of successive surveys provides a clear picture of the effects of the two spills and the early stages of recovery amongst some of these (figure 2). The response varied from the complete if temporary elimination of one species on each occasion (*Littorina obtusata*). or a steady decline

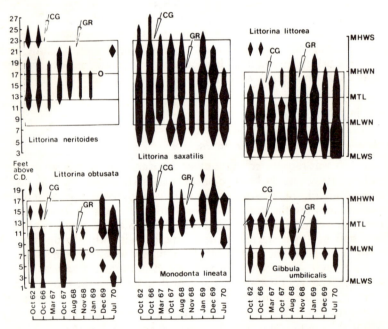

Figure 2 Distribution and abundance of some winkles and top-shells at Hazelbeach, Milford Haven, before and after two oil-spills (CG = tanker *Chryssi P. Goulandris*; GR = Gulf refinery) from surveys by A. Nelson-Smith and G. B. Crapp. Abundance has been graded as in figure 3. From Nelson-Smith (1972).

throughout the period (*L. neritoides*), to a slight increase in the abundance of another (*L. littorea*), presumably demonstrating a competitively advantageous resistance in the last. A further example of serious pollution occurring without warning but on a previously well-surveyed site was provided nearer the mouth of the Haven in 1973, when a

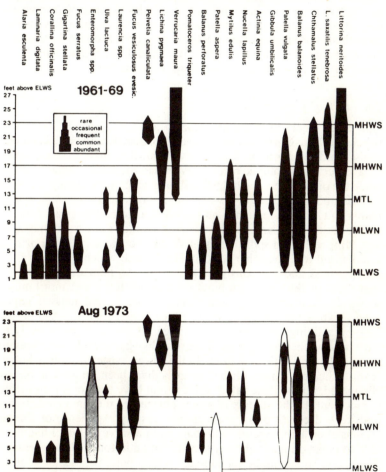

Figure 3 Distribution and abundance of common plants and animals of a shore in the mouth of Milford Haven during the years before, and shortly after, a locally serious gasoline spill. The pre-spill outlines for limpets *Patella vulgata* and *P. aspera* have been superimposed on the later survey to emphasize the reduction in their range and numbers. The consequent 'flush' of green algae is shown in stipple. From Nelson-Smith (1977).

products tanker was blown ashore in Lindsway Bay by a gale and large quantities of gasoline were spilt into strong surf; these conditions had the effect of emulsifying and retaining the gasoline within the small bay, although it might normally have been expected either to evaporate or drift away (Blackman *et al.*, 1973). This pollution narcotized large numbers of limpets, other gastropods, and mussels, rather than killing them outright, but many were subsequently eaten by seabirds before they could recover. Since limpets and winkles are largely responsible for controlling algal growth on shores of moderate exposure, by grazing away their sporelings shortly after they settle, this produced the expected 'flush' of annual green algae (figure 3), although the area affected is too small for this to have had long-lasting consequences.

The wreck of Torrey Canyon

Only a few weeks after the first Hazelbeach spill, *Torrey Canyon* was wrecked off the Cornish coast. Largely because of unexpected changes in the wind direction and hence the drift of the oil-slick, some shores became seriously polluted before surveys had been undertaken on them, while the oil failed to reach the few shores which were hastily surveyed in their undamaged state. It was quite impossible to determine which plants and animals remained on the worst affected shores until the oil had been dispersed, an operation which did much more harm than after any Milford Haven spill. I then surveyed two transects, one north of Land's End and the other west of The Lizard, which had been both heavily oiled and vigorously 'cleansed'. It was immediately obvious that most of the sedentary biota were unhealthy if not moribund—seaweeds were limp and discoloured, barnacles and mussels gaped widely, while the bodies of many limpets remained attached to the rock even though their shells had fallen off. Animals more likely to have been washed away, such as the snail-like gastropods, were present in much fewer numbers than might have been expected. Empty home-scars indicated where further limpets had been more completely eliminated, although not all of these would have been occupied before the incident. In the absence of a pre-spill survey, the main guide to the undisturbed condition of these shores was my experience of similar regions elsewhere, together with data from a survey which I had been able to make on a not entirely similar but an untouched shore east of The Lizard. It should be added that it was impossible, then or later, to carry out any field investigation of the effects of stranded *Torrey Canyon* oil alone because of the determination of the authorities to 'clean up' every oiled shore, however remote, by over-use of highly toxic dispersants. Figure 4 shows a comparison of immediately post-spill findings with surveys carried out six months later; the losses

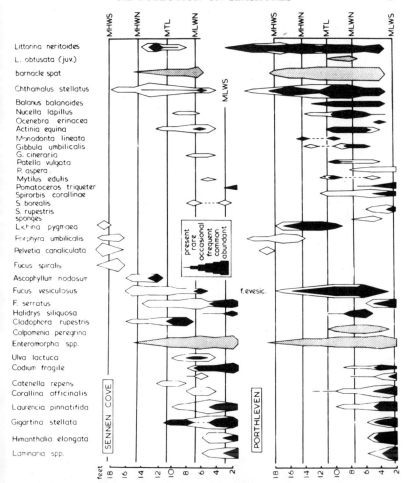

Figure 4 Distribution and abundance of surviving plants and animals of two shores in Cornwall a few days after (white) and six months after (black) the stranding of *Torrey Canyon* oil and its removal and toxic dispersants. Stippled histograms indicate new settlement of green algae and young sedentary animals. From Nelson-Smith (1968).

indicated occurred amongst biota which had already survived the initial effects of pollution and clean-up treatment, or were at least still present *in situ*. It also shows a marked 'green phase' and new settlements of barnacle spat, few of which survived beneath the vigorous algal growth. Because the damage was so widespread, the green algae were succeeded by the larger and long-lived fucoids whose dense cover persisted for several years before even an approximation to the previous balance was re-established (see Nelson-Smith, 1977).

Effects of a refinery effluent

A rather different use of shore surveys of this sort has been made around Little Wick, near the mouth of the Haven. Here, a refinery effluent of small volume is discharged at low-water level, unlike those

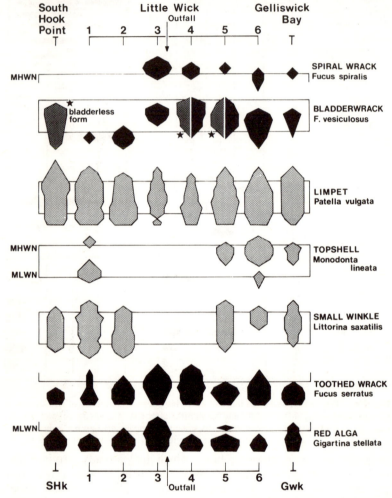

Figure 5 Distribution and abundance of the main seaweeds and grazers on shores to each side of an outfall of oily water from a Milford Haven refinery. Transects 1–6, added by G. B. Crapp, were established between the original ones at South Hook Point and Gelliswick Bay. High abundance grades were not used in assessing limpet density, so that the reduction in their numbers around the discharge point is not fully expressed. From Nelson-Smith (1975).

of the other installations which are released into deep water and strong currents near the jettyhead of their associated terminals. A photograph of Little Wick before the development of Milford Haven as an oil-port shows it to have been generally free of the larger seaweeds; ten years later, there were abundant growths of brown algae. The original transects at South Hook Point and Gelliswick lay too far to each side to show significant changes, so Crapp (1971a) surveyed six additional transects between them. Comparison of the distribution and abundance of limpets and other grazers on these transects with that of the larger algae (figure 5) shows that some constituent of the effluent is unfavourable to these animals. It cannot be claimed with certainty that the factor is oil, although this is consistently present in small quantities, because the discharge is largely of fresh water and also has other potentially damaging constituents, but the suppression of the limpet population in its vicinity is clearly associated with the denser growth of fucoid algae which, on some of the transects, are made up of an atypical mixture of the bladderless and normal forms of *Fucus vesiculosus*. If this outfall should ever be abandoned or relocated it would be most interesting to observe the establishment of a new balance between seaweeds and grazers.

Some seven years after the first rocky-shore surveys of Milford Haven, Crapp (1971b) sought evidence of the effects of the expanding oil industry and the occasional spillages resulting from its activities. The re-surveyed transects showed rather marked changes in the balance between the barnacle *Balanus balanoides* (which has a northerly distribution in the British Isles) and the southern species *Chthamalus stellatus*, which could reasonably be associated with the small but steady decline in sea temperatures noted during this period; a slight reduction in the abundance of the top-shell *Monodonta lineata* which can probably be assigned to the same change in climate; and a similar reduction in numbers of the topshore winkle *Littorina saxatilis tenebrosa*. Only the latter is a likely consequence of oil pollution. Subsequent surveys show mainly an increase in the density of cover by the green alga *Enteromorpha* at several sites (Baker, 1976); it is tempting to represent this as a probable consequence of the ill-effects of chronic low-level oil pollution or of repeated larger spillages, but such an interpretation would undoubtedly be an oversimplification.

Problems of interpretation

In an unpublished discussion which took place during the London meeting of the 'Challenger' Society early in 1968, J. R. Lewis commented that there had been an unusually heavy flush of green algae on the Yorkshire coast during the previous summer, coinciding with the one in Cornwall which followed *Torrey Canyon* oiling, but without any obvious unnatural cause; indeed, *Enteromorpha* often grows very well there following unusually heavy rains in the late spring. Although neither he nor any of the biologists investigating the aftermath of that tanker wreck were in any doubt that the 'green flush' recorded on Cornish shores was a consequence of the widespread mortality amongst grazing organisms, it is impossible to provide rigid proof that such enhanced growth would not have occurred in the absence of the pollution. Similarly, it is always possible that changes revealed by the comparison of before-and-after surveys may be wholly or partially due to the expression of some factor quite unrelated to the pollution incident under investigation. Interpretation of the data has to be regarded partly as a matter of experience. Extremes in seasonal or annual changes were reported by Lewis in 1977, when he was able to present numerical data on mussels, barnacles, and limpets collected regularly on an exposed and rather impoverished coast in the north-east over a period of ten years at upper and middle levels, and for five years on the lower shore (figures 6 and 7). Considered in the present context, the extremes are not quite as dramatic as they at first appear, since adult populations on the topmost site show considerable stability; sharp peaks in *Balanus* numbers are due to the settlement of larvae and subsequent high mortalities amongst the young barnacles, which are not normally enumerated at such a small size in the transect surveys described here (but see figure 4). In the middle shore, an even greater proportion of the barnacle population is made up of newly settled young, which here suffer predation by dog-whelks, *Nucella*, as well as suffering from physical conditions which are often unfavourable. Greater fluctuations amongst mussels reflect increases due to summer growth (note that the criterion used is percentage cover, not absolute numbers) or occasional heavy settlement, and decreases resulting from predation or storm dislodgment, mostly occurring in autumn and winter respectively. At low levels, there is even greater instability, apparently related mainly to a marked lack of diversity; the south-western shores with which I have been most concerned support a much greater variety of biota and hence demonstrate a much more stable state from year to year. Lewis also found greater stability on more sheltered sites.

Thus, although there is no necessity to cast undue doubt on the ability of single or infrequent transect surveys to demonstrate the effects of gross spillage or chronic local pollution on richly populated or well-sheltered shores, one wonders how easy it would be to interpret the results of two successive surveys in Robin Hood's Bay during periods when the lower shore was dominated by barnacles and mussels respec-tively, or when both were present in minimal numbers but limpets

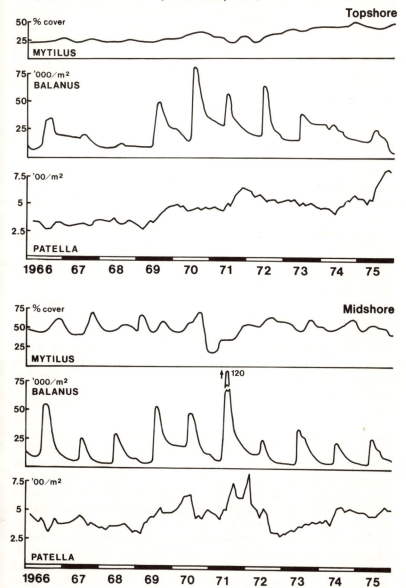

Figure 6 Fluctuations in the density of mussels, barnacles, and limpets measured at regular intervals on the Yorkshire coast for ten years at top- and mid-shore levels. Redrawn from Lewis (1977).

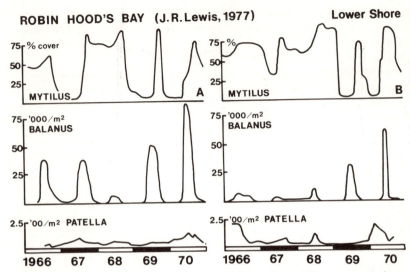

Figure 7 Fluctuations in the density of mussels, barnacles, and limpets measured at regular intervals on the Yorkshire coast for five years at two lower-shore locations. Redrawn from Lewis (1977).

reached a modest peak (compare, for example, figure 7*a* in mid-1967 and mid-1968 with figure 7*b* early in 1970). Comparison of the scales used in these diagrams with criteria given in table 1 illustrates why Lewis advised the introduction of higher abundance grades; even discounting obviously short-term recruitment peaks, the density of barnacles at upper and mid-shore levels rarely fell below a score of Abundant, while limpets at these levels passed into the Extremely Abundant category and mussels on the lower shore often attained the percentage cover which defines the grade of Superabundant. Such high densities are, however, associated with low diversity as commonly as is instability. It should also be borne in mind that young limpets, which spend their first few months in damp crevices and small pools, will have been included in Lewis's counts but ignored under the protocol of the belt-transect surveys. These regions would have been excluded from the area surveyed on the original transects, mainly so that the data collected could also be used in calculating an exposure grade for each shore (see below). In the light of experience and with the aid of its critics, now may be the time to modify the basic method further, at least when used in pollution monitoring, although preferably not so radically that fresh surveys would no longer be usefully comparable with earlier ones. The diagrams by which the data collected from belt-transect surveys have been presented here are quantitative in the sense that abundance grades were plotted proportionately throughout; but, as each diagram is a plane projection of vertically equidistant stations from a shore on which there are considerable variations in slope and topography, the total area of the histogram representing any organism will be proportional to overall numbers on the shore only when its zone is extremely narrow. From data drawn up in this way, it is possible to estimate losses after a pollution incident for which there are before-and-after surveys only at individual stations, not for the complete transect belt. An estimate for the whole of an affected shore could be attempted, although with no great numerical accuracy, if adequate details of the shore profile had also been recorded at the time of survey. However, the method was not intended for this purpose and the scientific value of such estimates is dubious.

Simplified methods

It is reasonable to assume that most of the ecological changes recorded in before-and-after transect surveys are brought about by the effects of pollution (or other changes in the quality of the local environment) on only a few key species. There is thus considerable attraction in the suggestion that only these key species need to be investigated; in the greater time available for such a simplified survey, actual counts could be made, or other data collected, which are more accurately quantifiable than assessments of abundance on a coarse scale and thus offer the possibility of more sensitive detection of adverse effects. I will not consider here the arguments for and against the usefulness of such 'indicator species', which have been discussed by Lewis (1976) amongst many others. However, the limpet *Patella vulgata* offers a relevant example. Limpets have long been recognized as key animals on shores of moderate exposure; the effects of any serious reduction in their numbers are both obvious and long-lasting. The consequences of serious spills in Milford Haven and Cornwall, summarized above, together with experimental work such as that by Dicks (1973), make it clear that limpets are very susceptible to pollution by hydrocarbons, whether in the form of crude oil, refined products, or dispersant solvents. In situations where the seaweed-grazer balance is not so obviously disturbed, such as around the Little Wick outfall, size-frequency analysis of measurements made after Geoffrey Crapp's transect surveys and reported by Baker (1976) indicate effects on the population structure. The largest limpets are to be found at the sites most affected by the oily effluent, where total numbers are fewest, and it seems that this is largely due to the failure of settling young to survive there. There was hope that the biotic state of any rocky shore might be determined by making such an analysis of its limpet population although, when this was attempted for a selection of shores around

Swansea Bay which are subjected to a variety of unfavourable natural and industrial influences, little if any pattern could be detected in the results (Nelson-Smith, 1974). Shell dimensions can readily be taken without disturbing the animals whereas data on their reproductive state (which can be regarded as another potentially valuable index) are more difficult to acquire, need to be taken at the appropriate time of year and demand the sacrifice of a sample of the population. It is interesting in this context to note that even the few large old limpets nearest to the Little Wick outfall contain ovaries whose weight falls within the normal range for their size.

It has long been recognized that the shell size, shape, and thickness in *Nucella lapillus* reflect the degree to which the shore on which it lives is exposed to strong winds and wave-action. John Crothers has made collections of these dog-whelks on the shores of south-west and southern Britain (1973, 1974, 1975) and from Portugal (1977) to Norway (not yet published). He has been able to relate their shell-shape mathematically with the exposure as expressed on a biologically-based scale devised by Ballantine (1961) and derivable from belt-survey data (Moyse and Nelson-Smith, 1963; Nelson-Smith, 1967). Unfavourable influences other than exposure can affect the shell-shape; Crothers (reporting in an unpublished paper to the 1978 Field Meeting of the Estuarine and Brackish-water Sciences Association) believes that the relationship with exposure can be expressed closely enough to permit the detection of chronic pollution, even at a low level, on any shore for which a reliable exposure grade can be derived. A field demonstration of this, on a reasonably large scale, is awaited with interest.

Limpets are of less importance on sheltered shores naturally dominated by algae and, perhaps, also on the most exposed sites. Dog-whelks may also be few in great shelter and absent from the most exposed shores. In each case, there is by no means universal agreement over the interpretation of the numerical data. Complete reliance on studies of such indicators thus appears to have its limitations, although each can no doubt be further refined. However, whenever available time and other circumstances permit, nothing would be lost by supporting a routine belt-transect survey with the more detailed investigation of these and other key species.

References

Addy, J., Baker, J. M., Dicks, B., and Levell, D. (1973) The intertidal biology of Sullom Voe, Shetland, and proposals for a biological monitoring scheme. *Annual Report of the Oil Pollution Research Unit, Orielton*, 1973, 47–71.

Ardré, F., Cabañas-Ruesgas, F., Fischer-Piette, E., and Seoane, J. (1958) Petite contribution á une monographie bionomique de la Ria de Vigo. *Bulletin du Institut Océanographique de Monaco*, **55**, 1–56.

Arnold, D. F. (1959) Report of work undertaken during tenure of a Research Fellowship in Marine Biology at Swansea, 1958–1959. London: The Development Commission (typescript).

Baker, J. M. (1976) *Marine Ecology and Oil Pollution*. Barking: Applied Science Publishers.

Ballantine, W. J. (1961) A biologically-defined exposure scale for the comparative description of rocky shores. *Field Studies*, 1 (3), 1–9.

Blackman, R. A. A., Baker, J. M., Jelly, J., and Reynard, S. (1973) The *Dona Marika* oil spill. *Marine Pollution Bulletin*, 4, 181–182.

Corlett, J. (1948) Rates of settlement and growth of the 'pile' fauna of the Mersey estuary. *Proceedings and Transactions of the Liverpool Biological Society*, 55, 2–28.

Crapp, G. B. (1971a) In *The Ecological Effects of Oil Pollution on Littoral Communities*, ed. Cowell, E. B. pp. 187–203. London: Institute of Petroleum.

Crapp, G. B. (1971b) In *The Ecological Effects of Oil Pollution on Littoral Communities*, ed. Cowell, E. B. pp. 102–113. London: Institute of Petroleum.

Crapp, G. B. (1973) The distribution and abundance of animals and plants on the rocky shores of Bantry Bay. *Irish Fisheries Investigations, series B*, 9, 1–35.

Crisp, D. J. (1958) The spread of *Elminius modestus* Darwin in north-west Europe. *Journal of the Marine Biological Association of the United Kingdom*, 37, 483–520.

Crisp, D. J. and Southward, A. J. (1958) The distribution of intertidal organisms along the coast of the English Channel. *Journal of the Marine Biological Association of the United Kingdom*, 37, 157–208.

Crothers, J. H. (1973) On variation in *Nucella lapillus* (L.): shell shape in populations from Pembrokeshire, South Wales. *Proceedings of the Malacological Society of London*, 40, 319–327.

Crothers, J. H. (1974) On variation in *Nucella lapillus* (L.): shell shape in populations from the Bristol Channel. *Proceedings of the Malacological Society of London*, 41, 157–170.

Crothers, J. H. (1975) On variation in *Nucella lapillus* (L.): shell shape in populations from the south coast of England. *Proceedings of the Malacological Society of London*, 41, 489–498.

Crothers, J. H. (1977) On variation in *Nucella lapillus* (L.): shell shape in populations towards the southern limit of its European range. *Journal of Mollusc Study*, 43, 181–188.

Dicks, B. (1973) Some effects of Kuwait crude oil on the limpet, *Patella vulgata*. *Environmental Pollution*, 5, 219–229.

Evans, R. G. (1947) The intertidal ecology of selected localities in the Plymouth neighbourhood. *Journal of the Marine Biological Association of the United Kingdom*, 27, 173–218.

Evans, R. G. (1949) The intertidal ecology of rocky shores in South Pembrokeshire. *Journal of Ecology*, 37, 120–139.

Fischer-Piette, E. and Seoane-Camba, J. (1962) Écologie de la ria-type; la Ria del Barquero. *Bulletin du Institut Océanographique de Monaco*, 58, 1–36.

Fischer-Piette, E. and Seoane-Camba, J. (1963) Examen écologique de la Ria de Camariñas. *Bulletin du Institut Océanographique de Monaco*, 61, 1–38.

Gabriel, P. L., Dias, N. S., and Nelson-Smith, A. (1975) Temporal changes in the plankton of an industrialized estuary. *Estuarine and Coastal Marine Science*, 3, 145–151.

Gilet, R. (1959) Water pollution in Marseille and its relation with flora and fauna. *Proceedings of the First International Conference on Waste Disposal in the Marine Environment*, 39–56.

Hillman, R. E. (1975) In *Fisheries and Energy Production*, ed. Saila, S. B. pp. 55–76. Lexington, Mass.: D. C. Heath/Lexington Books.

Jones, E. B. G. and Eltringham, S. K. (1971) *Marine Borers, Fungi and Fouling Organisms of Wood*. Paris: Organisation for Economic Co-operation and Development.

Kitching, J. A. (1950) The distribution of the littoral barnacle *Chthamalus stellatus* around the British Isles. *Nature, London*, **165**, 820.

Lewis, J. R. (1953) The ecology of rocky shores around Anglesey. *Proceedings of the Zoological Society of London*, **123**, 481–549.

Lewis, J. R. (1954) The ecology of exposed rocky shores of Caithness. *Transactions of the Royal Society of Edinburgh*, **62**, 695–723.

Lewis, J. R. (1964) *The Ecology of Rocky Shores*. London: English Universities Press.

Lewis, J. R. (1976) Long-term ecological surveillance: practical realities in the rocky littoral. *Oceanography and Marine Biology Annual Review*, **14**, 371–390.

Lewis, J. R. (1977) In *Biology of Benthic Organisms*, ed. Keegan, B. F., Ceidigh, P. O., and Boaden, P. J. S. pp. 417–423. Oxford: Pergamon.

Lysaght, A. M. (1941) The biology and trematode parasites of the gastropod *Littorina neritoides* L. on the Plymouth breakwater. *Journal of the Marine Biological Association of the United Kingdom*, **25**, 41–67.

Moyse, J. and Nelson-Smith, A. (1963) Zonation of animals and plants on rocky shores around Dale, Pembrokeshire. *Field Studies*, **1** (5), 1–31.

Naylor, G. L. (1930) Notes on the distribution of *Lichina confinis* and *L. pygmaea* in the Plymouth district. *Journal of the Marine Biological Association of the United Kingdom*, **16**, 909–918.

Nelson-Smith, A. (1965) Marine biology of Milford Haven: the physical environment. *Field Studies*, **2**, 155–188.

Nelson-Smith, A. (1967) Marine biology of Milford Haven: the distribution of littoral plants and animals. *Field Studies*, **2**, 435–477.

Nelson-Smith, A. (1968) In *The Biological Effects of Oil Pollution on Littoral Communities* (Supplement to Field Studies, vol. 2), ed. Carthy, J. D. and Arthur, D. R. pp. 73–80. London: Field Studies Council.

Nelson-Smith, A. (1970) Techniques: surveying rocky shores. *Fieldworker*, **1**, 50–52.

Nelson-Smith, A. (1972) *Oil Pollution and Marine Ecology*. London: Paul Elek.

Nelson-Smith, A. (1974) In *Report of the Working Party on Possible Pollution in Swansea Bay*, Vol. I, Chap. 2; Vol. II, Chap. 2. Cardiff: Welsh Office/HMSO.

Nelson-Smith, A. (1975) In *Petroleum and the Continental Shelf of North West Europe, 2 Environmental Protection*, ed. Cole, H. A. pp. 105–111. Barking: Applied Science Publishers.

Nelson-Smith, A. (1977) In *Recovery and Restoration of Damaged Ecosystems*, ed. Cairns, J., Dickson, K. L., and Herricks, E. E. pp. 191–207. Charlottesville: University Press of Virginia.

Nicholson, N. L. and Cimberg, R. L. (1971) In *Biological and Oceanographical Survey of the Santa Barbara Channel Oil Spill 1969–1970. I. Biology and Bacteriology*, ed. Straughan, D. pp. 325–399. Los Angeles: Allan Hancock Foundation, University of Southern California.

Percival, E. (1929) A report on the fauna of the estuaries of the R. Tamar and the R. Lynher. *Journal of the Marine Biological Association of the United Kingdom*, **16**, 81–108.

Purchon, R. D. (1957) The marine fauna of five stations on the northern shores of the Bristol Channel and Severn Estuary. *Proceedings of the Bristol Naturalists' Society*, **29**, 213–226.

Rattray, J. (1886) The distribution of the marine algae of the Firth of Forth. *Transactions of the Botanical Society of Edinburgh*, **16**, 420–466.

Smith, J. E. (ed.) (1968) '*Torrey Canyon*' *Pollution and Marine Life*. Cambridge: The University Press.

Southward, A. J. and Crisp, D. J. (1954) Recent changes in the distribution of the inter-tidal barnacles *Chthamalus stellatus* (Poli) and *Balanus balanoides* L. in the British Isles. *Journal of Animal Ecology*, **23**, 163–177.

Southward, A. J. and Orton, J. H. (1954) The effects of wave-action on the distribution

and numbers of the commoner plants and animals living in the Plymouth break-
water. *Journal of the Marine Biological Association of the United Kingdom*, **33**, 1–19.
Stephenson, T. A. and Stephenson, Anne (1972) *Life Between Tidemarks on Rocky
Shores*. San Francisco: W. H. Freeman.
Woods Hole Oceanographic Institution (1952).

Systematic surveys and monitoring in nearshore sublittoral areas using diving

KEITH HISCOCK

Field Studies Council Oil Pollution Research Unit,
Orielton Field Centre. Pembroke, Dyfed, Wales

Introduction

Until recently, systematic studies of sublittoral benthos have been mainly confined to areas which can be sampled or observed by the use of remotely operated equipment such as dredges, trawls, grabs, and television. In many nearshore shallow sublittoral areas these techniques cannot be used effectively particularly where quantitative records or very accurately located stations are required. In the rugged and variable topography of some nearshore sublittoral areas, free diving often offers the only means of making effective observations or collections.

The use of diving for studies of underwater populations has developed very slowly since the pioneer work of Kitching, Macan, and Gilson (1934) and even since SCUBA equipment first became widely available in the early 1950s. The use of diving in marine biology in British waters has accelerated in recent years and a description of available techniques and equipment has been prepared for publication by the present author. This paper is therefore concerned with the strategy of carrying out biological studies using diving.

Underwater around Britain the biologist will find a highly variable habitat with correspondingly varied populations of plants and animals. Shallow-water rocky areas where light intensity is high will generally be dominated by algae and, on the open coast, the zone is usually characterized by a kelp forest. This is referred to as the infralittoral zone. As the observer swims deeper, the algal cover thins out as light is attenuated and animals become predominant. This is the circalittoral zone. On coasts open to clear oceanic water the algal dominated zone extends to about 20 m, but in some turbid nearshore waters it extends to less than 5 m. Within the same depth, most of the differences in plant and animal populations at different sites are brought about by variation in the degree of exposure to wave action and tidal streams and by the stability of substrates present. The type of sediment present on the seabed is also largely governed by the degree of wave action and the strength of tidal flow, with vigorous water movements preventing

55

the deposition of or removing fine sediments. Sediment grade and the mobility of the sediment are of major importance to species living on and in the sediment.

Diving requires no special skill though good training is essential. The basic equipment used by the diver-biologist is usually the same as for the sports diver and the specialist items of sampling apparatus are generally not expensive nor complicated. The physiological limits of the diver, particularly in the cold turbid waters surrounding Britain, mean that efficiency and speed are essential in underwater studies. Safety also plays a large part in the planning of surveys and the Code of Practice for Scientific Diving (Underwater Association, 1974), together with any government-initiated safety regulations, is an important reference for any project leader.

Planning and carrying-out the work

The first requirement in planning all surveys is to understand what questions are being asked. Having established what the object of the survey is, the location of sampling sites and the techniques most appropriate to the requirements of the survey can be decided. At this stage, it is also necessary to consider the methods which will be used to analyse the data so that the information can be collected in a suitable form. The project director should also consider whether diving is really the most efficient or effective means of achieving the aims of the survey or whether remote observation or sampling could be used.

Practical aspects of working underwater

A minimum of two divers will be required in any survey with a third person if a boat is to be used. The divers either work as a pair or one acts as a standby diver. In some cases, it is most efficient for a diver to work alone; a partner can easily stir-up sediment so that work at one site becomes impossible, dislodge equipment, or distract the working diver when contact has to be ensured in poor visibility. A competent fieldworker can, on entering cold turbid water, become forgetful and incompetent. Furthermore, it is not often possible to sit in one spot and scan the study area to write a list of species. Neither can the worker be sure of remembering everything seen during a dive when there are many other things to be thought about than just the survey task.

The following factors should be considered at the planning stage:
1. Account should be taken of the experience of the divers—in diving, in species recognition, and in the work to be carried out. The absence of a general guide to the identification of sublittoral species and the lack of formal education at university in rocky sublittoral ecology in

particular will make the use of inexperienced personnel difficult. In surveys of species distribution, the worker should not be expected to learn recognition as he or she goes along and some days might be needed for training. The individual skills of each worker should be used to greatest effect.

2. The sea-water temperature and likely tolerance of workers to cold need to be taken into account. The efficiency of divers working in wet suits drops considerably at temperatures below about 12 °C, though this depends on experience, enthusiasm, and individual tolerance. Dry suits, though expensive, make comfortable, warm working conditions possible in cold water.

3. When operating in depths greater than 10 m, the time available on the seabed before decompression becomes necessary should be borne in mind. Also, nitrogen narcosis and anxiety make tasks more difficult with increasing depth.

4. The safety procedures appropriate to diving and boating need to be accounted for in planning.

5. Problems of access to sites for shore-based parties and of launching facilities for boat-based surveys should be considered.

6. The times and duration of slack water at proposed sites should be noted. Admiralty Coastal Pilots and Charts include information on slack water times.

7. The exposure of coasts to wave action should be taken into account: exposed coasts should be surveyed at the first opportunity, and each day's work made complete so that usable information is obtained even if winds prevent further work at that site.

8. The limitations which poor visibility and/or poor light will impose on the survey should be considered. Work becomes difficult in a horizontal visibility of less than about 3 m or where a torch is required. This is particularly the case where habitats or stations have to be located or where divers are working together.

9. The ease with which sites can be re-located where surveys are to be carried out at the same site over several days or where monitoring is the aim should be borne in mind. It might be appropriate to allow time for the laying of guide lines or other site marks.

10. Account should be taken of the variation in species abundance through the year and from year to year. The abundance and range of species of algae varies greatly from summer to winter. Surveys of several locations aimed at comparing differences in species composition should be carried out at or as near as possible the same time in order to minimize the possibility of seasonal changes providing the differences. In monitoring from year to year, surveys should be made at the same time each year as far as possible.

11. In monitoring studies, it is prudent to consider whether survey

will be possible at set intervals or whether flexibility will be required because of variable environmental conditions which might make diving or the required work impossible (for instance, turbidity or wave action).

12. An assessment should always be made of the time available to sort and identify collected material or transcribe data both during the survey and afterwards.

Range of species and habitats to be studied

The project director must determine the survey or sampling units to be investigated and the range of species which are to be included to satisfy the aims of the work.

Selection of survey or sampling units or the definition of the main units to be studied will be based on, for example:

1. The main biotic zones with depth—the infralittoral fringe, upper infralittoral, lower infralittoral, upper circalittoral, circalittoral.

2. Particular habitats or substrates. For example, whole *Laminaria hyperborea* plants, *L. hyperborea* holdfasts, upwardly inclined unbroken rock surfaces, sediment plains.

The range of species to be studied or sampled and the detail in which records are to be made must also be decided. Choice of the range of species includes:

1. 'All' species (from large widely dispersed species to small undergrowth species larger than 0.5 mm).

2. Prominent species easily recognized *in situ*.

3. Undergrowth species requiring collection for identification.

4. Key, characteristic, or 'indicator' species.

5. Species groups (e.g. foliaceous algae, kelp, bryozoa/hydrozoa).

6. Species within taxonomic groups (e.g. animals, algae, coelenterates, sponges).

7. Single species.

The detail in which records are to be made will be based on:

1. Quantitative samples or measurements (carefully collected random samples or measurements from known area expressed as density, percentage cover or weight).

2. Semi-quantitative samples or records (collections from roughly delimited area, estimates or rough counts of density or percentage cover).

3. Qualitative (presence/absence) records.

Location of sites

Surveys are likely to fall into two categories: extensive (primary) and intensive. The extensive survey is often aimed at determining the habitats and species present, their distribution in relation to environmental conditions, and the range of variation of sublittoral habitats and populations over a wide area (greater than about 5 km of coast). Intensive surveys may be concerned with the accurate tracing of gradients in species distribution and abundance in a small area, studies of the extent or frequency of occurrence of species or habitats, investigations of local distribution and ecology of a particular species, or following changes in species or populations of organisms over several years.

Extensive surveys will usually be carried out over a wide range of habitat types and at sites exposed to a variety of environmental conditions. The worker can easily select sites exposed to markedly different environmental conditions such as depth, wave action, and tidal stream velocity by careful reference to maps, charts, and coastal pilots. Such features as the extent of sublittoral bedrock are less easy to judge from charts. Within the main habitat or environmental categories, the sites which can be surveyed will often be selected on the basis of features such as ease of access and the presence of appropriate substrates. The selection is systematic but not random and is usually the best way to plan a survey aimed at describing the variety of habitats or populations present in an area. If random selection of sites is required (for instance if the data are required to assess the frequency of occurrence of particular habitats or populations) then, in a small area, all grid line intercepts of the coast can be used. In a large area some form of random selection can be carried out from all of the possible locations within each environmental category (this is the 'stratified random sampling' of terrestrial ecologists). The random selection of sites based on an objective classification of the coastal environment for primary surveys of shallow sublittoral zone is advocated by Earll (1977).

Intensive surveys often require reconnaissance before determining sampling intervals, the species to be included in the survey, the most suitable techniques, and the delineation of the area to be investigated. Some areas of seabed might be apparently exposed to similar environmental conditions for their whole length or area whilst in some studies it might be required to follow the effect of a physical or chemical gradient along the coast. Such situations might be suitable for survey at regularly spaced intervals or over a grid. However, particularly in the rocky sublittoral, the worker will almost certainly be faced with such problems as ease of access and, where samples are required from particular habitats, the extent of the relevant habitat. In such cases, sites will have to be spaced only as near as possible to the chosen distance (see for instance Moore, 1973). Distance between sites will depend in particular on the area to be studied, the gradient being studied, and the time available. In surveys of sites from shallow to deep water along a transect or swim line, studies at depth intervals or within the main zones would seem most appropriate. However, it should be remembered that change is rapid in the transition zones such as the infralittoral fringe and infralittoral-circalittoral transition area so that additional sites might be required.

Surveys aimed at describing the habitats or plant and animal populations present at underwater sites have to be carried out within a systematic framework if the results from different sites and from the records of different workers are to be comparable. Care must be taken to record only within certain depth zones and from certain topographical features or to make clear in the results the seabed features from which the species were recorded. The species included on checklists will depend on the range of plants and animals present in the area and the species-recognition ability of fieldworkers. For prominent easily recognized species, the fieldworker can expect to observe up to 60 species of animals and 50 species of algae in one dive. The site features recorded as a background to the biological work will depend on the aims of the survey. As an example of a site recording card, figure 1 shows an edge-punched card designed for the South-west Britain Sublittoral Survey where habitat description is a key part of the work.

Surveys of a particular species, community, or of apparently homogeneous habitats often require the collection of quantitative data for statistical analysis. Random sampling or recording is essential, but many sublittoral rocky areas are highly heterogeneous and a very large number of random samples would doubtless be required to provide representative samples of the whole site. Therefore, the type of substrate on which the samples are to be taken must be defined before the start of the survey.

Quantitative, semi-quantitative, or presence/absence data

The type of data collected will depend on the information needed to satisfy the aims of the survey, the time available for collection and processing of data, the methods of analysis to be used, and the level of accuracy which is possible. A common problem in studies of rocky areas is to find that the effort of collecting much of the data in a survey has been wasted because the methods used in analysis make detailed data redundant. Thus, counting all of the individuals of a species where each is solitary but estimating the percentage cover or weighing colonial species will necessitate comparing two different units of measurement and the worker is likely to have to create a scale of abundance and fit all of the accurate data into broad categories, making detailed counts redundant. Some benthic ecologists can happily count individual specimens and exclude from analysis the sponges, hydroids, and bryozoa which cannot be counted as individuals; this is a convenient but undesirable technique in describing whole populations, except where

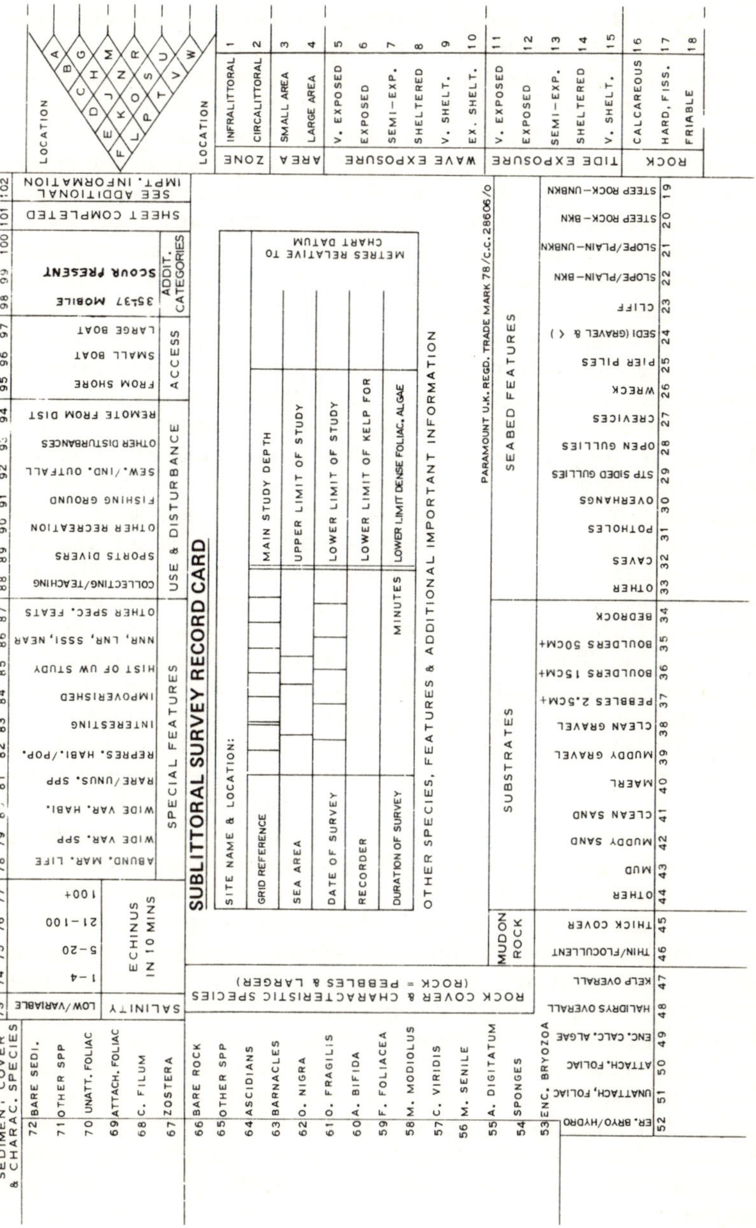

SUBLITTORAL SURVEY RECORD CARD

PARAMOUNT U.K. REGD. TRADE MARK 78/C.C.28606/0

SITE NAME & LOCATION:

- GRID REFERENCE
- SEA AREA
- DATE OF SURVEY
- RECORDER
- DURATION OF SURVEY

METRES RELATIVE TO CHART DATUM

- MAIN STUDY DEPTH
- UPPER LIMIT OF STUDY
- LOWER LIMIT OF STUDY
- LOWER LIMIT OF KELP FOR
- LOWER LIMIT DENSE FOLIAC. ALGAE
- MINUTES

OTHER SPECIES, FEATURES & ADDITIONAL IMPORTANT INFORMATION

LOCATION

A B G / C H M / D J N / E K O / F / Z R Q / S U / T V / W — LOCATION

ZONE / AREA
1 INFRALITTORAL
2 CIRCALITTORAL
3 SMALL AREA
4 LARGE AREA

WAVE EXPOSURE
5 V. EXPOSED
6 EXPOSED
7 SEMI-EXP.
8 SHELTERED
9 V. SHELT.
10 EX. SHELT.

TIDE EXPOSURE
11 V. EXPOSED
12 EXPOSED
13 SEMI-EXP.
14 SHELTERED
15 V. SHELT.

ROCK
16 CALCAREOUS
17 HARD, FISS.
18 FRIABLE

SEABED FEATURES
19 STEEP ROCK–UNBKN
20 STEEP ROCK–BKN
21 SLOPE/PLAIN–UNBKN
22 SLOPE/PLAIN–BKN
23 CLIFF
24 SEDI (GRAVEL & <)
25 PIER PILES
26 WRECK
27 CREVICES
28 OPEN GULLIES
29 STP SIDED GULLIES
30 OVERHANGS
31 POTHOLES
32 CAVES
33 OTHER

SUBSTRATES
34 BEDROCK
35 BOULDERS 50CM+
36 BOULDERS 15CM+
37 PEBBLES 2.5CM+
38 CLEAN GRAVEL
39 MUDDY GRAVEL
40 MAERL
41 CLEAN SAND
42 MUDDY SAND
43 MUD
44 OTHER

MUD ON ROCK
45 THICK COVER
46 THIN/FLOCCULENT

ROCK COVER & CHARACTERISTIC SPECIES (ROCK = PEBBLES & LARGER)
47 KELP OVERALL
48 HALIDRYS OVERALL
49 ENC. CALC. ALGAE
50 ATTACH. FOLIAC.
51 UNATTACH, FOLIAC.
52 ER. BRYO/HYDRO
53 ENC. BRYOZOA
54 SPONGES
55 A. DIGITATUM
56 M. SENILE
57 C. VIRIDIS
58 M. MODIOLUS
59 F. FOLIACEA
60 A. BIFIDA
61 O. FRAGILIS
62 O. NIGRA
63 BARNACLES
64 ASCIDIANS
65 OTHER SPP
66 BARE ROCK

SEDIMENT COVER & CHARAC. SPECIES
67 ZOSTERA
68 C. FILUM
69 ATTACH, FOLIAC
70 UNATT, FOLIAC
71 OTHER SPP
72 BARE SEDI,

SALINITY
73 LOW/VARIABLE

ECHINUS IN 10 MINS
74 1–4
75 5–20
76 21–100
77 100+

ABUND. MAR. LIFE
78 WIDE VAR. SPP
79 WIDE VAR. HABI.
80 RARE/UNUS. SPP
81 REPRES. HABI./POP.
82 INTERESTING
83 IMPOVERISHED

SPECIAL FEATURES
84 HIST OF UW STUDY
85 NNR, LNR, SSSI, NEAR
86 OTHER SPEC. FEATS

USE & DISTURBANCE
87 COLLECTING/TEACHING
88 SPORTS DIVERS
89 OTHER RECREATION
90 FISHING GROUND
91 SEW./IND. OUTFALL
92 OTHER DISTURBANCES
93 REMOTE FROM DIST

ACCESS
94 FROM SHORE
95, 96 SMALL BOAT
97 LARGE BOAT

ADDIT. CATEGORIES
98 MOBILE
99 SCOUR PRESENT
35137

100 SHEET COMPLETED
101, 102 SEE ADDITIONAL IMPT. INFORMATION

Figure 1 An edge-punched card used for recording site details. Three holes are used where it is required to record the feature as predominant, secondary, or minor. The card was designed for the Nature Conservancy Council South-west Britain Sublittoral Survey.

species which cannot be counted are rare. The use of an abundance scale in presenting or analysing results has two main advantages: it does not require accurate counting or weighing, and it can be compiled to reflect the real 'abundance' of a species by taking into account the maximum quantity in which that organism is likely to occur. An abundance scale for rocky sublittoral species is shown in figure 2.

Many species have a widespread distribution but are present in very different numbers from site to site. Recording actual numbers, weight, percentage cover, frequency of occurrence or abundance means that the ecological importance of each species related to the quantity present can be seen by the reader of a report or will be taken into account during analysis. Presence/absence records provide much less useful information for the comparison of sites. It can hardly be argued that a species should be given the same importance if it is present in densities of several thousand per square metre as in densities of one or two per square metre (as happens when the species is recorded as 'present'). However, for the purposes of some surveys aimed at identifying marked gradients, sites with a similar species composition or species with a similar distribution, presence/absence data may be adequate. This is particularly the case where a large number of species are recorded. Hoare and Hiscock (1974) found that the effects of a highly toxic effluent could be illustrated adequately by recording the position along the coast at which littoral and sublittoral species were first observed away from the effluent. Moore (1974) found that, in the analysis of data on holdfast fauna, quantitative information gave only a small increase in information relative to that given by qualitative species composition. Moore (1974) has discussed in detail the merits of using qualitative rather than quantitative information in the analysis of benthic data and concludes that 'the demarcation point beyond which the extra informativeness of quantitative data becomes progressively unnecessary would seem to be at about 30 species'. However, Eagle (1974), working on the fauna of mud and sand, notes that it was necessary to use quantitative data for samples which included up to 96 species to obtain ecologically meaningful results. Monitoring in particular would seem to require quantitative or semi-quantitative data rather than records of presence/absence although Moore (1974) suggests that 'there seems to be no *a priori* reason why satisfactory surveillance of an area from the point of view of recording change over a period of time could not be achieved by repetition at intervals of a primary qualitative survey'. My experience of studying seasonal and longer-term changes in rocky shore species indicates that most organisms are present throughout the year and from year to year but often in widely different numbers.

Species category	per station (ca 100 m²) 1-2	3-5	6-9	per 10 m² 1-2	3-5	6-9	per 1 m² 1-2	3-5	6-9	per 0.1 m² 1-2	3-5	6-9	per 0.01 m² or % cover 1-2	3-5	6-9	10-20	21-50	51-99	100+
Large solitary or clumped species present in generally small numbers (eg some echinoderms, massive sponges, hydroid clumps).	1	2	3	4	5	6	7	8	>										
Large solitary or clumped species present in generally large numbers (including kelp).	<	1	2	3	4	5	6	7	8	>									
Small but easily visible species (eg cup corals, anemones, some ascidians, large non-clumped hydroids).				<	1	2	3	4		5	6	7	8	>					
Small species sometimes in very large numbers (eg barnacles).					<	1	2	3		4	5		6	7				8	
Small species generally visible only in collected samples (eg annelids, small crustacea, small molluscs).							<	1		2	3	4	5	6	7		8	>	
Foliaceous algae and crustose species.				<	1		2	3		4	5		6	7	8				>

Figure 2 An abundance scale for converting field records to grades appropriate to each species recorded. Only examples of some species groups which would be included under each category are given. For each survey, the species included in each category must be listed. The grades are equivalent to notations as follows: 1, Very rare; 2, Rare; 3, Occasional; 4, Frequent; 5, Common; 6, Very common; 7, Abundant; 8, Very Abundant.

Quantitative survey and sampling

Quantitative survey is carried out using single quadrats of an appropriate size or belt-transects established by a line laid over the seabed
or by rolling a quadrat over along the seabed. Quantitative sampling
of undergrowth fauna and flora is carried out using a suction sampler
(Hiscock and Hoare, 1973). Many of the methods used for determining
the location of survey stations or sampling areas in studies of terrestrial
vegetation can be applied to rocky sublittoral plant and animal populations and texts such as Grieg-Smith (1964), Shimwell (1971), and
Kershaw (1977) are useful references.

The location of sampling stations can be determined by: (*a*) the
subjective selection of a 'representative' area, (*b*) the random location
of quadrats within predetermined habitat categories or apparently
homogeneous populations, (*c*) the completely random dispersion of
quadrats at the site, or (*d*) the use of a series of contiguous quadrats.
Random sampling is essential if statistical methods are to be applied
to the data but wholly random sampling is inappropriate in rocky
sublittoral areas where rugged topography and a rapid change in
population-type with depth leads to a high degree of heterogeneity. In
rocky areas, it is essential to define the seabed features and depth in
which counts are to be made or from which samples are to be collected.
Thus, a diver might be instructed to locate an area of 'upwardly inclined
bedrock at least 3 m above the nearest sand and at a depth of 20 m
below Chart Datum'. If the seabed is largely unbroken, a tape measure
might be easily laid and sampling points located from random number
tables. There is rarely time to carry out the difficult task of laying a rope
or tape grid underwater. On an unbroken seabed, random sampling
by quadrats may often be achieved by dropping the quadrat onto the
area from a height above the seabed where the surface cannot be seen.
However, it is usually the case that the worker is instructed to use the
first area he can find which satisfies the definition of the sampling area
and onto which the quadrat will fit.

In carrying out any quantitative work, the minimum survey or
sampling area required to obtain a statistically acceptable mean density for particular species or, in presence/absence data, a list of all
common species, must be considered. If possible, the size of the area
should be established before the main study commences though this is
not always possible, particularly where several man-days sorting and
identification are required for each sample. A species-area curve such
as that shown in figure 3 is usually used to determine minimum sampling
area. Drew (1971) suggests that a suitable point to select as a minimum
sampling area for presence/absence data is that at which a 100 per cent
increase in area yields only 10 per cent or perhaps 5 per cent more

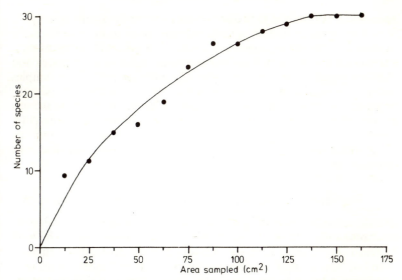

Figure 3 A species-area curve used to determine minimum sampling area (from a *Petroglossum nicaeense* association near Marseille). Re-drawn from Boudouresque (1971).

species. Some organisms cover large areas of seabed or may be sparsely distributed and therefore require a minimum sampling area of several tens or hundreds of square metres. Others will require only fractions of a square metre. No tests are known to have been carried out on the large widely dispersed organisms to establish minimum areas. However, the abundance scale shown in figure 2 gives an indication of the area likely to be occupied by different groups of species. It is useful to consider here the few measurements of minimum sampling areas which have been made on sublittoral rocks. Boudouresque (1971) found that a sampled area of 100 cm^2 was adequate to list almost all of the algae present in shallow shaded areas, while 250 cm^2 was needed in the coralligenous zone of the circalittoral in the Mediterranean. Larkum, Drew, and Crossett (1967) found that 400 cm^2 was the minimum sampling area required for algal vegetation off Malta, but note that samples of over 1 m^2 were needed to establish significant differences ($p < 0.05$) between dominant species in similar vegetation at 30 and 45 m. Few tests have been carried out to establish a minimum sampling area for rocky sublittoral animal populations. From the data collected by Hiscock (1976) it has been found that, in 0.5 m^2 samples from three sites at Lough Ine, as many as half of the total of 50, 55, and 83 species recorded at the three sites occurred only in one or other of the separate 0.25 m^2 samples. For non-rare species (more than one or two individuals per sample or more than a few sprigs), about a quarter of the total species were present in either one or the other of the separate samples. This represents up to a 20 per cent increase in the number of non-rare species per 100 per cent increase in area sampled, and indicates that, for these populations, 0.5 m² was not an adequate sampling area and that 1 m² might have been preferable. Unpublished records for samples taken all around Lundy suggest a similar increase of about 20 per cent of all species present on doubling the sample area from 0.25 to 0.5 m². However, it is clear that the small amount of information presently available is inadequate to allow the worker planning a sampling programme to decide on the sample or survey area before fieldwork commences.

Describing significant changes with time in rocky sublittoral populations

Studies aimed at describing changes with time in marine populations are commonly referred to as 'monitoring' or 'surveillance'. The term 'monitoring' is used here without any attempt to further define its meaning and is used as synonymous with 'surveillance'. There is at present considerable discussion of the aims of monitoring and of the techniques most suitable to satisfy those aims. It is often the case that, when the aims of a particular exercise are clear, the most suitable techniques (if any) to carry out the work become obvious. Three main problems exist in carrying out monitoring on the rocky seabed in British coastal waters using diving: (*a*) the great heterogeneity of rocky sublittoral populations resulting from a rugged underwater topography, (*b*) the difficulty of refinding monitoring sites in conditions of low visibility, and (*c*) the limited time available for underwater work.

Only gross changes can be assessed by comparing results from descriptive surveys at the same site. Recording from marked areas is more accurate and can sometimes be greatly assisted by the use of photography. Lundälv (1971) has used stereo pairs of photographs to describe seasonal and longer term changes on sublittoral rocks off Sweden. However, stereophotography has no advantage where only simple counts or measures of percentage cover are required. Photography cannot be used where some species are obscured by other organisms. Measures of percentage cover or frequency for encrusting species *in situ* will require the use of point quadrats or cross-wire frames which will doubtless be difficult to handle under the influence of wave action or tidal streams. However, if this method is used, it will be essential to establish the number of strikes required to assess accurately the percentage cover or frequency.

Lewis (1978) advocates the use of key species only in monitoring but, without knowing what species are 'key', initial work at least will usually be concerned with the inclusion of all non-rare species. Selecting a small range of species for survey will lead to difficulty in interpretation of results if, at a later survey, a previously unrecorded species is dominant. Counting only key species or common species will make the underwater task much more rapid and should be considered carefully in relation to the aims of the work.

It must be realized that the use of quadrats at marked sites will provide information on changes within those quadrats which may not necessarily be representative of all of the surrounding area.

Environmental variables

A knowledge of the environmental variables likely to affect the distribution and abundance of species will be required at the planning stage of a survey, particularly if it is intended to study the effect of one physical or chemical factor and reduce the variation in others by careful selection of sites.

The biologist carrying out surveys can rarely be concerned with the direct measurement of physical conditions which in many cases vary through the year. However, the data will often be available from oceanographic studies carried out in the same area or extrapolated from more widespread reports.

The following factors are likely to be important:

1. *Exposure to wave action and tidal streams.* The strength of water movements is extremely important in determining the grade of sedimentary substrates present and the plants and animals present in both sediments and on rock. Published data on wind velocity, direction, and frequency together with standard oceanographic formulae will enable the grading of a shore according to wave action at the surface. However, wave action is attenuated with depth and calculation of seabed oscillatory velocity under all wave conditions through the year is time-consuming. Records of surface tidal flow can be obtained from Admiralty Coastal Pilots and Charts. Attenuation of tidal flow velocity occurs with depth but to a much lesser extent than wave action, so comparison of sites based on surface records of tidal streams is acceptable in most cases. The calculation of wave-induced oscillatory flow velocity and tidal flow velocity on the seabed is described by Hiscock (1976). Figure 4 shows graphs which can be used to grade sites according to their exposure to or shelter from wave action and tidal streams. The scales cannot accurately represent actual velocities of flow on an uneven seabed but provide the best available method of comparing sites.

2. *Light* is attenuated with depth and the spectral composition of light changes with depth. The amount of light reaching the seabed will vary through the year in relation to day-length, cloud-cover, angle of

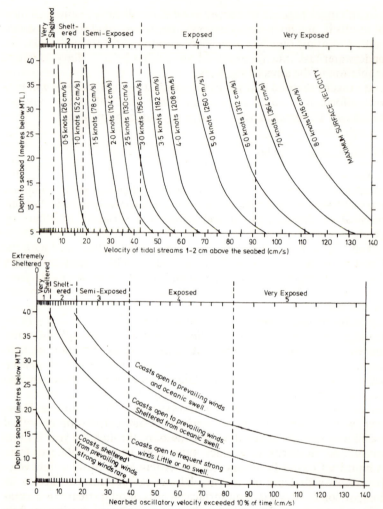

Figure 4 An exposure scale for open areas of sublittoral seabed. The velocity of tidal streams at about 1 to 2 cm above the seabed and the near-bed oscillatory velocity exceeded for 10 per cent of the time is given to indicate the actual strength of water movement to which the scale applied. From Hiscock (1976).

the sun, and, most markedly in many locations, with irregular variations in suspended sediment concentrations. Fluctuation in light intensity related to suspended sediment levels makes the obtaining of a representative value for light penetration very difficult. Shading by topographical features or organisms near to the seabed is also important. Weinberg (1976) reviews the importance of submarine daylight to benthic organ-

isms and describes equipment to record the intensity of light in the sea
in an ecologically meaningful way. Drew (1971) describes direct-
reading and recording photometers. Kain (1971) describes apparatus
used to monitor underwater light intensity throughout the year. Apart
from electronic instruments, Secchi disc recordings are useful as a
rough means of recording light penetration. Also, the maximum downward extent of the kelp forest or of algal domination gives a good
indication of light conditions on open seabed throughout the year
(though subject to influences such as grazing by sea urchins and where
light intensity is only very important in determining depth distribution
at the time of algal settlement).

3. *Depth*. Records of depth should be related to a standard level and
Chart Datum (near to Lowest Astronomical Tide Level, LAT) is
generally used. Height of sea-level above CD at the time of a dive can
be calculated from data supplied in Admiralty Tide Tables. Depth
gauges are often inaccurate and must be calibrated in a pressure
chamber or in the sea against a vertical shot-line.

4. *Suspended sediment concentration* controls light penetration and
also affects the amount of silt deposited on sublittoral substrates.
Moore (1977) reviews the effect of inorganic particulate suspensions
on marine animals. The amount of sediment in the water column
varies throughout the year but is very different from area to area. For
instance, Moore (1973) records a range of 2–7 mg/l through one
year at Millport in the Firth of Clyde and a range of 22–307 mg/l at
Flamborough Head on the north-east coast of England. Thus, a figure
which represents the suspended sediment regime at a particular location can be obtained only from measurements taken throughout the
year. Moore (1972) describes methods for measuring the concentration
of suspended solids and for recording the deposition of silt.

5. *The annual temperature range* is important in determining the
species present at a particular location, both through their tolerance
to high and low temperatures and the effect of temperature on reproduction. Temperature regime throughout the year can be taken into
account with accuracy only if recordings are available for several
years from sources such as marine laboratories, field stations, power
stations, or dock authorities. The Conseil Permanant International
pour l'Exploration de la Mer (CPIEM) (1962) record annual variations
in surface temperature for British waters, and their data are usually
adequate for open coastal waters.

6. *Salinity differences* from site to site and through the year are
generally slight on the open coast. CPIEM (1962) record annual variation for open coast areas around Britain. The importance of salinity
level and variation has been described by many authors (see for instance
Perkins, 1974).

7. *Geographical location* is generally important to the widespread distribution of species. In very extensive surveys, the sampling sites might extend over different water masses or temperature regimes and some species might reach the limits of their geographical range within the study area. If it is intended to correlate location within the study area with the distribution of a species, a measure of site location along a one-dimensional gradient is required. Hiscock (1976) used distance along the coast from certain possible points of division of water masses around the island of Anglesey to correlate species distribution with location.

If it is intended to record the location of sites for general reference, it is usual to give latitude and longitude or a National Grid map reference. Biological recording schemes use various systems for recording location, and are described in Heath and Scott (1974). If it is intended to record location so that exactly the same site can be visited in future years, photographs, sketch maps and permanent markers on the shore should be used.

8. *Substrate type* is particularly important to species living in sediments. Mobile substrates will not allow colonization by sessile species whilst shingles, cobbles, and boulders may hold a limited fauna and flora which is destroyed during storms. The fauna of bedrock will be affected by topography and, to a lesser extent, by the geology of the rock and whether the rock is easily attacked by boring species, has crevices, etc. Substrate type can be categorized in the manner described later for *in situ* surveys. Methods for the accurate description of sediments are described by Buchanan (1971).

9. *Predator or dominant cover* species will affect the other organisms present and should be noted.

Conclusion

It is hoped that this paper will be used as a basis for the planning of surveys and monitoring in nearshore areas using diving techniques. Whilst my opinions on survey strategy will not be acceptable to many and will not cover every type of survey fieldworkers may be required to carry out, it is felt valuable to conclude with a list of recommendations.

Surveys

1. The aims of the survey should be clearly understood.
2. In planning a survey, the project should be discussed with persons familiar with the area, searches of the literature carried out, and inspections made of maps and charts to discover the range of habitats likely to be encountered, the species likely to be found, and any special

problems such as access to sites which will affect the sampling programme.

3. Decisions must be made as to whether remote or *in situ* observation/sampling is appropriate and to what extent samples rather than observations will be required. Remote observation (television) and sampling (dredges) may be used for reconnaissance or semi-quantitative survey over extensive areas of level seabed in depths greater than 30 m. Diving techniques may be used for reconnaissance, semi-quantitative survey, and quantitative sampling in nearshore rocky areas, and remote sampling (grabs) for quantitative survey of sediments. Diving techniques should be used for quantitative sampling of sediments where precision is required in the location of stations or where samples are required from the full depth of inhabited sediment.

4. The survey team should be selected on the basis of the skills and experience required in species recognition, equipment operation, and diving ability.

5. Recording sheets, check lists, and instruction sheets must be prepared for use during the survey.

6. If the area has not already been studied, a reconnaissance should be carried out for several days to establish an inflexible framework for systematic surveys including the addition of species to check lists, deciding site location, deciding depth intervals or distances between stations to include the main zones, and standardizing all aspects of the survey.

7. Each member of the team should understand all aspects of the survey procedure and be able to complete check lists accurately.

8. To ensure that the full range of habitats or populations are described, sites should be selected at locations where environmental conditions appear distinctly different. To ensure that sufficient data are collected to enable mapping or to provide data to calculate the extent of habitats/populations or their frequency of occurrence, sites should be included at regularly spaced but randomly located positions (e.g. OS grid lines).

9. The main biotic zones should be included at steeply sloping sites by placing stations at appropriate depth intervals. On gradual slopes or plains, stations should be placed at depth intervals and, where the depth intervals are more than 100 m apart, at intermediate locations.

10. Spot dives or samples should be supplemented by swim lines to map boundaries of substrates or populations.

11. On the seabed, a check list should be used to ensure that each category or species is searched for. A tape recorder provides the best method for the collection of large amounts of data. The actual density or percentage cover of species should be recorded, and the area of each depth interval searched until satisfied that no further important substrates, habitat-types, or species are present. Care needs to be taken to

record only from the station depth and to qualify records of species occurrence where present only in or on micro-habitats such as over-hangs, kelp stipes, or crevices.

12. The recording sheets should be completed from field notes as soon as possible after diving.

13. Over extensive areas of level seabed where detailed data are not required, a towed sledge should be used or a drift dive with a telephone or through-water communication to the surface to provide information on the range of substrates or species present.

14. If quantitative samples are required from particular stations an air-lift suction sampler, which is suitable for the substrate being sampled, should be used. If samples from sublittoral bedrock are being compared, the type of substrate to be sampled should be carefully defined. For random sampling in a homogeneous sediment or rock plain, a tape measure can be laid and random number tables used to locate the sampling station. Alternatively, the quadrat can be dropped from a height out of visible range of the seabed. On broken irregular rock surfaces, the diver should place the sampling quadrat on the nearest surface which satisfies the definition of the type of substrate/habitat to be sampled. On sediment, at least 0.1 m^2 should be collected for des- criptive surveys of the most common species or at least 0.5 m^2 if de-tailed species lists are required. Samples should be collected to a depth of 15 cm on mud to 50 cm on gravel. On rock, at least 0.2 m^2 should be collected for descriptive surveys of the most common species and at least 1 m^2 if detailed species lists are required.

15. In situations where samples are required and the substrate is of loose-lying stones, rocks should be taken from within a delimited area using sample bags attached to a buoyed line.

16. If kelp holdfasts or whole kelp plants are used as a sampling unit, only mature plants of *Laminaria hyperborea* should be collected.

17. For measuring the density of large widely dispersed species, a 30 or 50 m weighted line is laid over the seabed and the survey team swims along each side of the line counting all of the individuals of the species being studied which are present under a metre rule held at right angles to the line.

18. For the illustration of habitats, substrates, or species, it is important to adopt a standard procedure with regard to the angle and distance between the camera and subject. The same light source should be used throughout. For illustrations of species, the same size reference frame should be used throughout to enable comparison of scale.

Monitoring rocky sublittoral populations

1. A reconnaissance of the area should be carried out to establish the

range of variation in populations present, to list the species present, and to consider particular difficulties of the area for the establishment of monitoring sites.

2. Tests should be made to establish minimum sampling or recording areas required to provide sufficiently precise data to satisfy the aims of the monitoring programme.

3. Permanent markers should be established for lines and/or quadrats. If necessary, permanent guidelines or grids can be laid to enable rapid re-location of stations.

4. Stations should be surveyed or photographed.

5. Sites should be re-surveyed at as near as possible regular intervals or at the same time of year using exactly the same techniques with the same degree of thoroughness as the initial survey.

Acknowledgements

Much of this paper was initially prepared as part of work commissioned by the Nature Conservancy Council to make recommendations on the most appropriate methods of carrying out surveys in nearshore sublittoral areas. I am grateful to my colleagues in the Field Studies Council Oil Pollution Research Unit for comments and criticism of the manuscript.

References

Boudouresque, C. F. (1971) Méthodes d'étude quantitative du benthos (en particular du phytobenthos). *Tethys*, **3**, 79–104.

Buchanan, J. B. (1971) Measurement of physical and chemical environment. Sediments. In *Methods for the Study of Marine Benthos*, ed. Holme, N. A. and McIntyre, A. D. pp. 30–52. Oxford: Blackwell Scientific.

Conseil Permanent International pour l'Exploration de la Mer: Service Hydrographique (1962) *Mean monthly temperature and salinity of the surface waters of the North Sea and adjacent waters from 1903 to 1954*. Charlottenlund Slot. CPIEM.

Drew, E. A. (1971) Botany. In *Underwater Science*, ed. Woods, J. D. and Lythgoe, J. N. pp. 25–68. London: Oxford University Press.

Eagle, R. A. (1974) Benthic studies around the North Wirral long sea sewage outfall. University of Wales: PhD thesis.

Earll, R. (1977) A methodology for primary surveys of the shallow sublittoral zone. In *Progress in Underwater Science*, ed. Hiscock, K. and Baume, A. D. pp. 47–63. London: Pentech Press.

Grieg-Smith, P. (1964) *Quantitative Plant Ecology*. London: Butterworth.

Heath, J. and Scott, D. (1974) *Instructions for Recorders*. Published by the Biological Records Centre, Monks Wood, Huntingdon.

Hiscock, K. (1976) The influence of water movement on the ecology of sublittoral rocky areas. University of Wales, PhD thesis.

Hiscock, K. and Hoare, R. (1973) A portable suction sampler for rock epibiota. *Helgoländer Wissenschaftliche Meeresuntersuchungen*, **25**, 35–38.

Hoare, R. and Hiscock, K. (1974) An ecological survey of the rocky coast adjacent to the effluent of a bromine extraction plant. *Estuarine and Coastal Marine Science*, **2**, 329–348.

Kain, J. M. (1971) Continuous recording of underwater light in relation to *Laminaria* distribution. In *Fourth European Marine Biology Symposium*, ed. Crisp, D. J. pp. 335–346. London: Cambridge University Press.

Kershaw, K. A. (1977) *Quantitative and Dynamic Plant Ecology*. London: Edward Arnold.

Kitching, J. A., Macan, T. T., and Gilson, H. C. (1934) Studies in sublittoral ecology. I. A submarine gulley in Wembury Bay, South Devon. *Journal of the Marine Biological Association of the United Kingdom*, **19**, 677–705.

Larkum, A. W. D., Drew, E. A., and Crossett, R. N. (1967) The vertical distribution of attached marine algae in Malta. *Journal of Ecology*, **55**, 361–371.

Lewis, J. R. (1978) The implications of community structure for benthic monitoring studies. *Marine Pollution Bulletin*, **9**, 64–67.

Lundälv, T. (1971) Quantitative studies on rocky-bottom biocoenoses by underwater photogrammetry. A methodological study. *Thalassia Jugoslavia*, **7**, 201–208.

Moore, P. G. (1972) Particulate matter in the sublittoral zone of an exposed coast and its ecological significance with special reference to the fauna inhabiting kelp holdfasts. *Journal of Experimental Marine Biology and Ecology*, **10**, 59–80.

Moore, P. G. (1973) The kelp fauna of north-east Britain. I. Introduction and the physical environment. *Journal of Experimental Marine Biology and Ecology*, **13**, 97–125.

Moore, P. G. (1974) The kelp fauna of north-east Britain. III. Qualitative and quantitative ordination, and the utility of a multivariate approach. *Journal of Experimental Marine Biology and Ecology*, **16**, 257–300.

Moore, P. G. (1977) Inorganic particulate suspensions in the sea and their effects on marine animals. *Oceanography and Marine Biology Annual Review*, **15**, 225–263.

Perkins, E. J. (1974) *The Biology of Estuaries and Coastal Waters*. London: Academic Press.

Shimwell, D. W. (1971) *The Description and Classification of Vegetation*. London: Sidgwick and Jackson.

Underwater Association (1974) *Underwater Association Code of Practice for Scientific Diving*. London: Natural Environment Research Council.

Weinberg, S. (1976) Submarine daylight and ecology. *Marine Biology*, **37**, 291–304.

Monitoring with deep submersibles

GILBERT T. ROWE

Woods Hole Oceanographic Institution, Massachusetts, USA

Introduction

The open ocean, because it appears so vast and far removed, has often been used as a recipient of urban and industrial wastes. Unfortunately, it does not have an infinite volume and now many local seas, basins, and estuaries have suffered from indiscriminate dumping. The very deep sea, because it is tranquil, cold, and seemingly of little ecological relevance, has been suggested as a possible repository for many of the more insipid materials we need to dispose of in the sea. But because we do not know the global importance of the deep-sea, we cannot mindlessly proceed with such activities. If life as we enjoy it now in shallow productive waters is vitally dependent on processes in the deep ocean, we must make certain that we know the consequences of using the deep-sea for waste disposal. To do this, monitoring is called for, but the great depths present formidable barriers to simple monitoring programmes.

One possible approach to monitoring great depths is the use of Deep Submergence Research Vessels (DSRVs). The Woods Hole Oceanographic Institution (WHOI) has operated DSRV *Alvin* since 1964 (Sharp and Shumaker, 1977) and it has been used in a variety of disciplines. Among them, increasingly, is the investigation of the effects of pollutants on the deep-sea biota and environment. I would like to describe the history of *Alvin*, its major accomplishments, and specifically how it is now being used in deep-sea environmental quality programmes.

The Alvin programme

Alvin, designed principally by Allyn Vine at WHOI, was built between 1962 and 1964 by Litton Industries, in the USA. Manned dives to 1800 m had been completed by the end of 1965. *Alvin* is 7.5 m long and 4 m high (figure 1). An interior sphere holds all controls, life support systems, and three persons (a pilot and two scientists). Batteries to power propulsion motors, the mechanical manipulator, lights, and cameras are outside the sphere, but enclosed in the superstructure

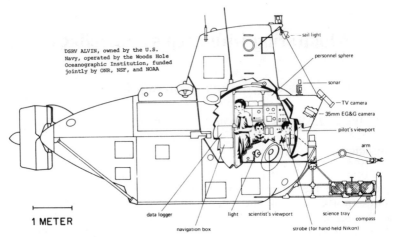

DSRV ALVIN, owned by the U.S. Navy, operated by the Woods Hole Oceanographic Institution, funded jointly by ONR, NSF, and NOAA

sail light
personnel sphere
sonar
TV camera
35mm EG&G camera
pilot's viewport
arm

1 METER

data logger   light   scientist's viewport   science tray   compass
navigation box   strobe (for hand-held Nikon)

Figure 1   DSRV *Alvin.*

fabricated from syntactic foam and fibre glass. The original sphere of steel was replaced in 1973 with one of titanium, increasing the maximum depth from 1800 m to 4000 m (12 400 ft).

Alvin's first spectacular success was recovery, with DSRV *Aluminaut*, of the hydrogen bomb lost off the coast of Spain. Early dives by natural scientists in 1966 to 1967 were mostly geological exploration and observation, including the first actual views of submarine canyons. In 1968 biologists gained greater use, observing Right whales, deep-living lobster populations, benthos, plankton, and observation and specific identification of species of fishes causing the deep sound-scattering layer.

In October, 1968, near tragedy struck the incipient programme. On a routine launch, cables used to lower the vessel down into the water, between the tender ship *Lulu*'s two hulls, broke, dropping *Alvin* down abruptly into the sea. It bobbed to the surface, allowing the pilot and observers to scramble free and swim to safety. But relief was ephemeral, because *Alvin* rolled over, filled with water, and sunk to a depth of 1400 m. Though no life was lost, this event obviously placed the US academic DSRV programme in jeopardy and the wave of enthusiasm and support for the programme was radically dampened for a while.

In August, 1969, ten months later, *Alvin* was recovered by the USNS *Mizar*, with help from DSRV *Aluminaut*, again. On recovery, interest turned to the degree of corrosion, biological fouling, and decay suffered by all the gear. To everyone's surprise, the crew's lunch, consisting of sandwiches, an apple, and thermos bottles of coffee and bouillon, had not decayed after ten months' exposure to deep-sea water. This serendipidous experiment implied that microbial activity in the deep-sea is very low and led WHOI biologist H. W. Jannasch and his associates to initiate a continuing programme to investigate this possibility (Jannasch and Wirsen, 1972; Jannasch *et al.*, 1971; Jannasch and Eimhjellen, 1972; Jannasch, Wirsen, and Winget, 1973; *et al.*).

When *Alvin* was rebuilt in May, 1971, a group of biologists at WHOI, believing the mounting evidence that biological processes in the deep-sea are slow, initiated the 'permanent bottom station' concept. We presumed that understanding and measuring the slow rates of biological processes in the deep ecosystem would require long-term observations in locales where considerable information already existed about the composition of the biota. To do this, we established permanent stations in the north-west Atlantic where experimentation could be initiated. The first of these, called DOS I (figure 2), was at the base of the continental slope at a depth of 6000 ft or 1850 m. This original location was

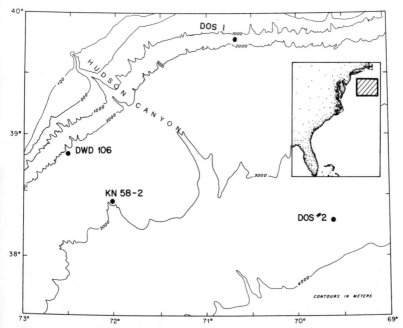

Figure 2 Map of the north-west Atlantic, showing permanent bottom stations and dump sites investigated by *Alvin*.

chosen not only because it was *Alvin*'s maximum depth, but because it was close to Woods Hole and was on the 'Bermuda Transect', sampled intensively by Howard Sanders' group of benthic ecologists. Among the studies accomplished at DOS I was the measurement of total sediment and bottom community oxygen consumption measured with polarographic O_2 electrodes in bell jar-like chambers (Smith and Teal, 1973). It was found that O_2 demand averaged about 0.5 cm^3 O_2 m^{-2} hr^{-1}, or several orders of magnitude below shallow water sediments, supporting the previous findings of Jannasch's group that deep-sea microbial metabolism is very low. Other stations were put in the Tongue of the Ocean off the Bahamas, in Hudson Submarine Canyon, and the Gulf of Maine, but only DOS I has been revisited consistently over a number of years. The replacement of *Alvin's* steel sphere pressure hull by one of titanium, between November, 1972 and June, 1973, increasing maximum depth to 4000 m (12 400 ft), permitted the addition of another important station, DOS II, again on the Bermuda Transect and at the new maximum depth to which the DSRV was allowed to go (figure 2).

A number of varied studies have utilized the bottom stations with *Alvin*. Ruth Turner has supplemented her on-going research on the ecology of wood-boring bivalve molluscs, and found results that seem to be exceptions to the generalizations that growth and substrate utilization rates are slow. Not only has she found that wood is infested and destroyed quickly in the deep-sea (Turner, 1973), but there is much evidence that the faecal matter produced by the borers has a broad impact on the surrounding community (Turner, in prepn). J. F. Grassle (1977), by deploying small boxes of mud free of animals at several of the stations, has found that rates of succession and growth are slow when no new source of organic matter is introduced at the same time. A two-ton bale of shredded urban refuse, placed at DOS I, in August, 1972, had not been radically changed when revisited two years later (Rowe and Clifford, 1975). Jannasch's group, using a variety of methods, has expanded its studies to include more general aspects of organic matter cycling (Jannasch and Wirsen, 1977), including *in situ* incubation of organic substrates, and deploying very large portions in the form of carcasses to compare microbial activities with utilization by fishes and invertebrates. Most of the ecological studies, including use of a sediment trap (Wiebe, Boyd, and Winget, 1976) at the station in the Bahamas, were related to the cycling of organic matter and the rates of biological processes, rather than the quality of the environment and monitoring. The National Oceanic and Atmospheric Administration (NOAA) and the Environmental Protection Agency (EPA) began, in 1974, the surveillance and monitoring of two dump sites off the east coast of the United States, and *Alvin* has played a big role in the base-line studies of these sites so far.

The dump sites are DWD106, formerly a CHASE site, used by the Department of Defence to dispose of munitions (CHASE being the acronym for 'Cut Holes and Sink 'Em'), and a low-level radioactive waste disposal site, the two areas being side by side and southeast of New York City, in the middle of the mid-Atlantic Bight (figure 2).

DWD106 is at present used for the offshore disposal of 'toxic' liquid industrial wastes whereas the 'Rad Site' is no longer receiving wastes, presumably. The studies of DWD106, with depths of 1300 to 2700 m, began in 1974 with surveys of the bottom fauna, using *Alvin* mostly, and conventional surface vessels, to record the physical and chemical composition of the water column.

Surveying the composition, standing stocks of the benthic fauna, and sediments required development of some special approaches. Large organisms, termed 'megabenthos', were quantified by making photographic transects with an 'odometer' to precisely determine distances across the bottom. The odometer is a bicycle wheel that is pushed ahead of *Alvin* along the bottom. A magnetic switch on the wheel records each revolution equivalent to 1 m distance travelled on a counter (digital) within the sphere. Observers, capable of taking 1500 photographs with *Alvin*'s cameras, record photograph numbers over each of the transects. Back in the laboratory, or in this case the laboratory converted to screening room, animals are counted per photograph per distance over the bottom. The area of each photograph is always the same as long as *Alvin* remains in contact with the bottom on the survey transects.

In front of the submersible is a large wicker wire basket into which are set all of *Alvin*'s sample equipment utilized with its mechanical arm. Most important to this work has been the tube-coring device and small box-corers (Rowe and Clifford, 1973) which were used at the origin and end of each photographic transect. The tube-corers, 10 cm in diameter, have been used for grain size, organic matter and dissolved inorganic pore-water nutrient analyses. The box-corers have been used for quantifying the invertebrate infaunal organisms. With these forms of sampling we have been able to quantify in one geographical area many of the size-classes of organisms previously only sparsely sampled at far-distant locales.

In 1974 the DWD 106 programme was directed by Robert Dill, in 1975 by Dan Cohen, and in 1976 by Murton Ingham, all of NOAA. Studies were also made during this period of similar baseline parameters in the 'Rad Site', under the direction of Robert Dyer of the EPA. After the 'stock assessment' of baseline studies of 1975, just described, we initiated at WHOI, in association with Ingham at NOAA, an investigation of the rates of processes at DWD 106 that could be related both to the flux of organic matter and to the flux and impact of pollutants at DWD 106.

At the end of the 1975 studies we concluded that the conditions at DWD 106, on the bottom at least, were no different from those which would exist if dumping did not occur there. Only the munitions, scattered widely over the area, marked it as a dump site. This might be logical, one might suppose, if only liquids are disposed of over great depths. But we proposed that several mechanisms might function to transport pollutants to deep water or to retain and accumulate them at density (temperature and salinity) discontinuities in the water column. While we believed we had documented conditions in the dump site, we realized we knew little about the cycling of matter in the ecosystem, and so our 1976 efforts were directed solely at measuring important rates of processes which might function in pollutant transfer.

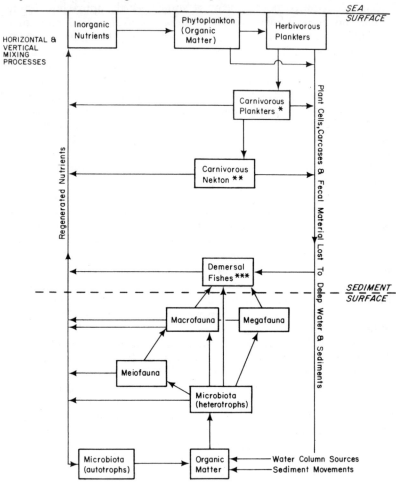

*   May or may not be diurnal migrators
**   May or may not be in schools, may or may not be invertebrates
***   May migrate off bottom

Figure 3   Box model of a deep ecosystem.

The rates we attempted to measure can be represented schematically as in figure 3, for the entire water column and the bottom. Interdependent state variables (all the squares), many measured previously, are linked by the fluxes of matter (the arrows) we were attempting to measure.

While many of the state variables and fluxes were measured with conventional surface-vessel techniques, many new approaches have recently been developed for submersible use. and these were utilized at DWD 106. Large quantities of the salp *Salpa aspera* had been observed in slope water in 1975, and we believed they could process particulate matter formed by Dupont waste into rapidly settling faecal pellets. To catch the mid-water salps and other delicate organisms, that by pellet production or vertical migration could move pollutants into great depths, a large slurp gun was built for use on *Alvin* by G. R. Harbison and Lawrence Madin. To catch particulate matter near the bottom, an array (figure 4) of sediment traps was deployed in the region that would be most likely to receive the pelletized particles (Rowe and Gardner, submitted). It has a release mechanism and trap covers that were operated manually by *Alvin*. Bell jar respirometers, used by K. L. Smith, were deployed to measure benthic metabolism. The question of sediment resuspension was approached with several new methods. To estimate bottom current shear forces, short (1 m) arrays of dye pellets at 20 cm intervals were implanted in the bottom and the rate of movement of turbulent streams away from the arrays, documented with 16 mm cinematography, was used to quantify water velocities at the sediment-water interface (Cacchione, Rowe, and Malahoff, 1978). Resistance of the sediment to such shear (or its shear strength) was determined with a small 'fall cone penetrometer' (figure 5). Time-lapse photography was used over an eight-day period to assess erosion of artificial tracks made in the bottom with small plaster-of-Paris animals, pulled in front of the camera by *Alvin* to make tracks and trails in the bottom.

Results of studies at DWD 106

Oxygen demand at the DWD 106 station was $0.46 \, cm^3 \, m^{-2} \, hr^{-1}$ (Smith, in press), and the rate of sedimentation of organic carbon was $6.3 \, g \, m^{-2} \, yr^{-1}$ (Rowe and Gardner, submitted). This can be put together, along with burial loss, as a budget of the flux of organic carbon (converting O_2 demand to CO_2 production from organic matter), as represented in figure 6 (Rowe and Gardner, submitted). Sedimentation from pelagic sources, as well as downslope gravity movements, is much greater at DWD 106, than offshore, we have concluded. On the bottom, current velocities are very low ($\sim 5 \, cm/s$) and shear strength of the

GILBERT T. ROWE

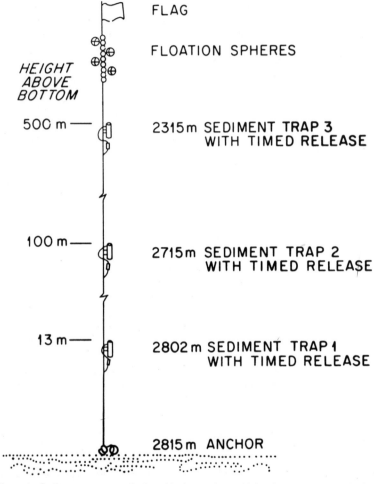

Figure 4 Sediment trap array deployed in the north-west Atlantic to measure particulate flux near the bottom (Rowe and Gardner, submitted).

sediments is very low (the mud is soft), indicating that this is an area of deposition without much erosion once the sediment is deposited. In other words, pollutants arriving here will stay. All the time-lapse studies confirm that this is a 'tranquil' area, rather than an active area, making it ideal if we are concerned with keeping a pollutant in a defined loca-

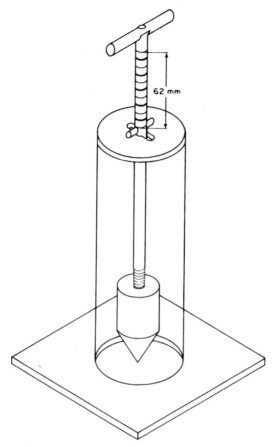

Figure 5 Fall-cone penetrometer used to measure shear strength of surface sediments from *Alvin*.

tion. Midwater studies with *Alvin* were successful in catching salps, but salps were not nearly as abundant in 1976 as in 1975 and their impact on pollutant transfer was negligible, as far as we could tell. Acoustic assessment (Orr, submitted) and dye studies of barges (Kohn and Rowe, submitted) indicated that the predominant wastes forming particulate matter formed concentrated but vertically narrow layers on density discontinuities in the water column.

The use of *Alvin* in the Rad Site by EPA has paralleled ours in DWD 106. That study has had the added need, however, of recovering old containers of waste, and they have done this successfully with *Alvin* (Dyer, 1976).

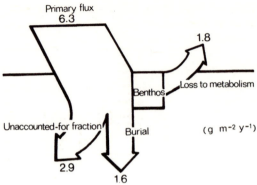

Figure 6 Organic carbon budget for DWD 106 off New Jersey, USA, summer 1976.

Conclusions on Alvin as a monitoring tool

DSRV *Alvin*, as demonstrated by the accomplishments summarized above, has some definite advantages, for the environmentalist, over conventional techniques. Few of the above studies could be done without a dexterous, strong manipulator, and the ability of the scientists to see things as they are happening. *Alvin*'s geological accomplishments (not summarized, but extensive) depend as well on extensive photography and on-bottom precise navigation. For experimental manipulations of many sorts, even though one can imagine remote free-vehicle or ship-tethered instrumentation to accomplish a task, it has been *Alvin*, with manipulator and real-time observation, that has repeatedly achieved the key results.

The need for ocean monitoring, however, is widespread, and there are only a few submersibles. Each requires pilots and observers to enter an alien environment, and so attached to each dive is an element of risk not found in conventional oceanography. DSRVs also, up to the present, have required special tending vessels and large teams of highly skilled technicians to operate them, meaning they are very expensive to use.

In most situations when *Alvin* was used in environmental studies, or large-scale geophysics, conventional vessels accompanied *Alvin* and *Lulu* on the expedition and were used continuously to collect more 'routine' data. Presently it is our intention, mostly because of cost and limited diving time available, to use *Alvin*, or any DSRV for that matter, to complete difficult experiments of a nature basic to deep-sea biology, which can be followed by shipboard monitoring programmes. As a result there are presently no *routine* environmental quality programmes in which *Alvin* participates. These are all done by surface vessels. *Alvin* continues exploring basic environmental quality questions, but is ready, with a suite of instruments, if and when needed for difficult surveillance, manipulation, and sample recovery.

References

Cacchione, D. A., Rowe, G. T., and Malahoff, A. (1978) Sediment processes controlled by bottom currents and faunal activity in Lowel Hudson Submarine Canyon. In *Sedimentation in Submarine Canyons, Fans and Trenches*, ed. Stanley, D. and Kelling, G. pp. 42–50. Stroudsburg: Dowden, Hutchinson and Ross.

Dyer, R. S. (1976) Environmental surveys of two deep sea radioactive waste disposal sites using submersibles. IAEA Paper SM-207/65. IAEA symposium, Vienna, Austria.

Grassle, J. F. (1977) Slow recolonisation of deep-sea sediment. *Nature*, **265** (5595), 618–619.

Jannasch, H. W. and Wirsen, C. O. (1977) Microbial life in the deep sea. *Scientific American*, **236** (6), 42–52.

Jannasch, H. W. and Eimhjellen, K. (1972) Studies of the biodegradation of organic materials in the deep sea. In *Marine Pollution and Sea Life*, FAO Conference on Marine Pollution, ed. Ruivo, M. London.

Jannasch, H. W. and Wirsen, C. O. (1972) ALVIN and the sandwich. *Oceanus*, **16**, 20–22.

Jannasch, H. W. and Wirsen, C. O. (1972) Deep-sea microorganisms: *in situ* response to nutrient enrichment. *Science*, **180** (4068), 641–643.

Jannasch, H. W., Eimhjellen, K., Wirsen, C. O., and Farmanfarmaian, A. (1971) Microbial degradation of organic matter in the deep sea. *Science*, **171** (3972), 672–675.

Jannasch, H. W., Wirsen, C. O., and Winget, C. L. (1973) A bacteriological pressure-retaining deep-sea sampler and culture vessel. *Deep-Sea Research*, **20** (7), 661–664.

Kohn, B. and Rowe, G. (submitted) Dispersion of two liquid industrial wastes dumped at Deep Water Dumpsite 106 off the coast of New Jersey, USA.

Orr, M. (in press) Acoustic detection of the particulate phase of industrial chemical waste released at DWD 106. *Journal of Geophysical Research*.

Rowe, G. and Clifford, C. H. (1975) A study of the ecological effects of solid waste disposal in the deep sea. Final Report to New England Regional Commission.

Rowe, G. T. and Clifford, C. H. (1973) Modifications of the Birge-Ekman box corer for use with SCUBA or deep submergence research vessels. *Limnology and Oceanography*, **18** (1), 172–175.

Rowe, G. T. and Gardner, W. P. (submitted) Sedimentation rates in the slope water of the northwest Atlantic Ocean measured directly with sediment traps.

Sharp, A. and Shumaker, L. A. (1977) DSRV ALVIN: a review of accomplishments. WHOI Technical Report 76–114.

Smith, K. L. and Teal, J. M. (1973) Deep sea benthic community respiration: an *in situ* study at 1850 meters. *Science*, **179** (4070), 282–283.

Smith, K. L., Jr. (in press) Benthic community respiration in the N. W. Atlantic: *in situ* measurements from 40 to 5200 meters. *Marine Biology*.

Turner, R. (1973) Wood-boring Bivalves, opportunistic species in the deep sea. *Science*, **180**, 1377–1379.

Wiebe, P. H., Boyd, S. H., and Winget, C. L. (1976) Particulate matter sinking to the deep-sea floor at 2000 m in the Tongue of the Ocean, Bahamas with a description of a new sedimentation trap. *Journal of Marine Research*, **34** (3), 341–354.

Continuous plankton records: monitoring the plankton of the North Atlantic and the North Sea

Institute for Marine Environmental Research,
Plymouth, England

The sampling problem

Synoptic surveys of the plankton of the North Atlantic Ocean and the North Sea have been carried out by regular sampling at monthly intervals since the beginning of 1948. The sampler used in the survey is the Continuous Plankton Recorder (Hardy, 1939). The organization of the survey has been described by Glover (1967) and various aspects of the methodology of sampling have been considered by Rae (1952) and Colebrook (1960, 1975a, 1975b).

Very briefly, Continuous Plankton Recorders are towed, by merchant ships and Ocean Weather ships, along a number of standard routes. The routes in operation in 1977 are shown in figure 1. Tows are made along each route at monthly intervals, subject to weather and the availability of ships.

The Plankton Recorders are towed at a standard depth of 10 m, and water is admitted through an aperture of 12.5 mm square. The plankton is filtered through a slowly moving band of bolting silk (24 meshes/cm). It is held in place by a second band of silk, the double band being wound continuously onto a storage spool in a tank containing formalin. About 10 cm of silk are wound across the water tunnel for every 10 miles of tow, during which about 3 m^3 of water are filtered. When the Plankton Recorders are returned to the laboratory the roll of silk is unwound and divided into sections each representing 10 miles of tow. For most records, alternate ten-mile sections are examined, the plankton being identified and counted.

With the exception of minor modifications to improve reliability and range, the design and method of deployment of the Recorders and the method of counting the zooplankton have not been changed over the last 30 years. The method of counting the phytoplankton was changed in 1958 to improve resolution (Colebrook, 1960).

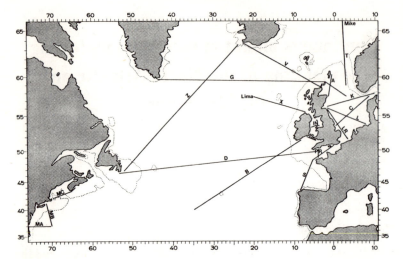

Figure 1 A chart of the North Atlantic showing the tracks sampled regularly by Continuous Plankton Recorders during 1977.

The survey provides a set of data consisting of counts of species (or higher taxonomic categories in instances where identification is difficult) at monthly intervals at 20-mile spacing along each route. These data can be used to provide information about seasonal, geographical, and annual changes in the abundance of the plankton.

It is well known that plankton data are extremely variable. The Recorder sample is a cylinder of water 12.5 mm in diameter and 10 miles long and clearly such a configuration plays a significant role in confounding small-scale spatial variability. But, even so, the best statistical model of populations as sampled during a single tow is a Negative Binomial distribution with a value for the exponent of 0.8 (Colebrook, 1975b). This gives a linear relationship between mean and standard deviation with a slope of just over 1 (figure 2b) and probability distributions of catches for varying population means as shown in figure 2a. The most efficient way of obtaining information about populations with such distributions is to take averages of relatively small counts of organisms from large numbers of samples (see, for example, Taylor, 1953). This concept provides the basis for our approach to obtaining information from the survey data, with the inevitable compromises imposed by the logistics of the survey and limitation in the man-power available for the analysis of samples.

The description problem

The difficulty lies in establishing criteria for the assessment of the

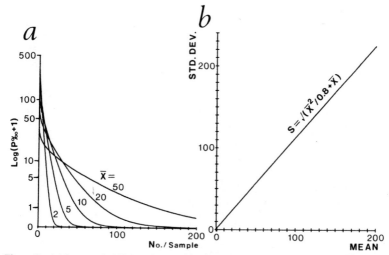

Figure 2 (a) Log probabilities of sample size for negative binomial distributions with an exponent of 0.8 and means of 2, 5, 10, 20, and 50. (b) A graph of standard deviation against mean for negative binomial distributions with an exponent of 0.8.

reality of descriptions of variations in the abundance of the plankton. From a computer simulation of the survey, providing opportunities for 'replication' it was shown that, provided averages were based on a sufficient number of samples, the data can provide adequate descriptions of events for about 50 of the more abundant species and other taxa of zooplankton and phytoplankton (Colebrook, 1975b).

The survey yields data that can be considered in terms of a multi-dimensional system with coordinates relating to species, areas, months, and years, providing a variety of two-dimensional subsets capable of analysis by standard multivariate techniques. It has been shown that, in many instances, there are sensible, systematic fluctuations in the variables and generally logical patterns of relationship between variables giving good, albeit circumstantial, evidence for the reality of the observed variability (Colebrook, 1978).

There are, of course, areas of uncertainty. Figure 3 shows graphs and a contour diagram of variations in the abundance of Chaetognatha in the south-eastern North Sea. Graph *a* shows the long-term mean seasonal variation. Each point is an average of well over 200 samples, and the seasonal cycle is similar to that shown by Chaetognaths in other areas. Graph *b* shows the year-to-year changes in abundance for the period 1951 to 1976. Each point is an average of about 100 samples. The fluctuations are similar to those for Chaetognaths in other areas of the North Sea and also to those for several other zooplankton taxa in the same area. There is no reason to believe that graphs *a* and *b* do not represent real events.

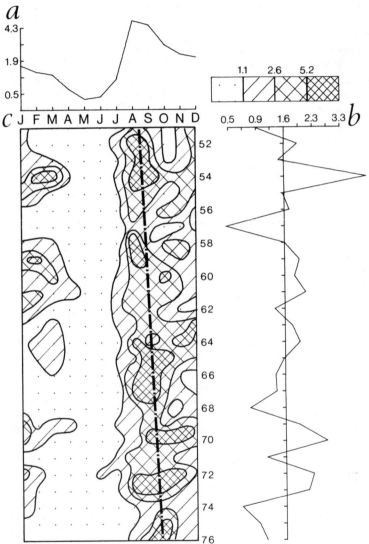

Figure 3 Diagrams illustrating fluctuations in the standing stock of Chaetognatha, as numbers per 3 m³, in the south-eastern North Sea (area D1 in the chart shown in figure 4): (a) Long-term seasonal averages; (b) Annual averages for 1951 to 1976; (c) A contour diagram of numbers in each month from January 1951 to December 1976. A key to the contour levels is given.

The contour diagram shows the seasonal variation in each year. Each point is an average of about ten samples. It would appear that there has been a slight shift in the timing of the seasonal maximum by about a month over the 25-year period, as indicated by the guideline in figure 3c. There is, however, an element of uncertainty about the reality of this apparent trend. Similar timing shifts have been observed for other taxa in other areas and, in some instances the addition of further years of data has necessitated modification of the initial interpretation. In addition, although figure 3b does not show any marked trends in abundance, the abundance of Chaetognaths in the central North Sea has shown a steady decline over the period and it may be that the apparent shift in timing is no more than a facet of this.

The diagrams presented in figure 3 are one example of 1231 such sets that are available, relating to different areas and to various taxa of zooplankton and phytoplankton. Each of the data sets is updated every year and graphs and contour diagrams like those in figure 3 are produced as a matter of routine. Clearly, there are considerable problems in assimilating and presenting this large mass of information. As part of the routine data processing, principal components analyses are used to produce relatively condensed descriptions of the main aspects of the seasonal, geographical, and annual changes in abundance.

In relation to monitoring, the main interest lies in the description of year-to-year changes. It is useful, however, to take a brief look at the main seasonal and geographical patterns in the plankton to provide the context within which annual changes take place and also because they were used as the proving ground for most of the analytical methods subsequently used in the study of annual changes.

Figures 4 and 5 show graphs for each of the first principal components of the seasonal variations in abundance of the more abundant taxa in each of the areas shown in the key chart given in figure 4. These components typically account for about 80 per cent of the variability within each data set. In each diagram the species have been ranked with spring species at the top and autumn species at the bottom.

For the zooplankton the rank makes sense in terms of a food web in which zooplankton eats phytoplankton and zooplankton eats zooplankton. There is a preponderance of filter feeders at the top of the list, with an increasing proportion of large-food feeders towards the bottom and with strict carnivores scattered through the rank. For the phytoplankton there is a discontinuity in the timing of the seasonal maxima, separating species involved in the spring bloom from the others. Most, but not all, of the spring species are small, while the summer and autumn species are all relatively large and contain all the relatively slow growing dinoflagellates.

For both phytoplankton and zooplankton there are taxa with peaks in abundance from March through to October and clearly there is scope for differences between years due to good or bad seasons within years.

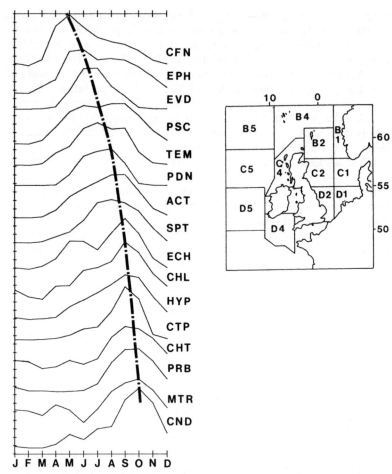

Figure 4 Graphs of the first principal components of the seasonal variations in abundance for 16 taxa of zooplankton based on data for each of the 12 areas shown in the key chart. On the *x*-axis each interval represents one standard deviation. A smoothed guide line linking the seasonal maximum of each taxon was added by eye. The taxa are: CFN, *Calanus finmarchicus*; EPH, Euphausiacea; EVD, *Evadne* spp.; PSC, *Pseudocalanus elongatus*; TEM, *Temora longicornis*; PDN, *Podon* spp.; ACT, *Acartia clausii*; SPT, *Spiratella retroversa*; ECH, *Euchaeta norvegica*; CHL, *Calanus helgolandicus*; HYP, Hyperiidea; CTP, *Centropages typicus*; CHT, Chaetognatha; PRB, *Pleuromamma robusta*; MTR, *Metridia lucens*; CND, *Candacia armata*.

Figure 6 contains charts of the first and second principal components of the geographical distribution of 43 taxa of zooplankton. In both components there are taxa whose distributions are inversely related to the patterns as presented in the charts. There are, for example, several

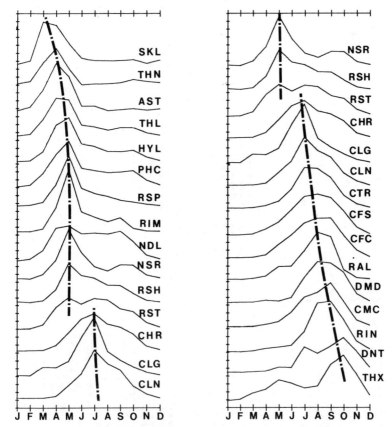

Figure 5 As figure 4 for 24 taxa of phytoplankton. The bottom six graphs in the first column are repeated at the top of the second column. The taxa are: SKL, *Skeletonema costatum*; THN, *Thalassionema nitzschioides*; AST, *Asterionella japonica*; THL *Thalassiosira* spp.; HYL, *Chaetoceros* (Hyalochaete) spp.; PHC, *Chaetoceros* (Phaeoceros) spp.; RSP, *Rhizosolenia hebetata semispina*; RIM, *Rhizosolenia alata inermis*; NDL, *Nitzschia delicatissima*; NSR, *Nitzschia seriata*; RSH, *Rhizosolenia imbricata shrubsolei*; RST, *Rhizosolenia styliformis*; CHR, *Ceratium horridum*; CLG, *Ceratium longipes*; CLN, *Ceratium lineatum*; CTR, *Ceratium tripos*; CFS, *Ceratium fusus*; CFC, *Ceratium furca*; RAL, *Rhizosolenia alata alata*; DMD, *Dactyliosolen mediterraneus*; CMC, *Ceratium macroceros*; RIN, *Rhizosolenia alata indica*; DNT, *Dactyliosolen antarcticus*; THX, *Thalassiothrix longissima*.

which occur in the North Sea and over the shelf areas but not in the open ocean (Oceanographic Laboratory, Edinburgh, 1973).

The distributions shown in figure 6 can be interpreted in terms of the basic topography and hydrography of the area (Colebrook, 1978). As with the seasonal variations in abundance there is clearly scope for differentiation from year to year dependent on the geographical locations of species.

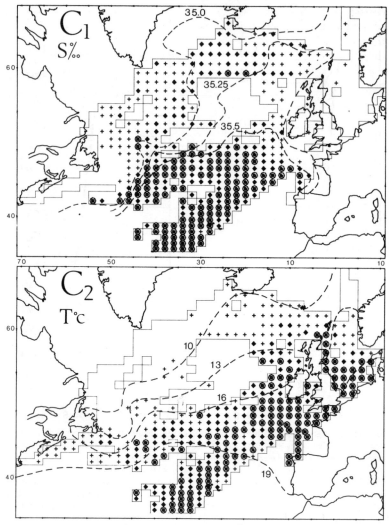

Figure 6 Charts of the first and second principal components of the geographical distributions of taxa of zooplankton. Values are represented by graded symbols:

| highest quarter | circle | third quarter | cross |
| second quarter | diamond | lowest quarter | blank |

The first component chart also shows surface isohalines in the range 35 to 36°/00 and the second component chart shows surface isotherms at 3° intervals for the month of August.

Colebrook (1978) has presented a detailed account of the annual fluctuations in the abundance of zooplankton in the north-eastern Atlantic and the North Sea for the period from 1948 to 1975. There is considerable coherence between the annual fluctuations in abundance of the various taxa in the different areas and this suggests strongly that the patterns of annual fluctuation in abundance are real; it also provides the means by which relatively simple descriptions can be produced of the main events over the last three decades. These are summarized in figures 7 and 8 which contain graphs of principal components with, superimposed on them, smoothed versions to emphasize long-term trends. Figure 7 contains graphs of the first principal components of

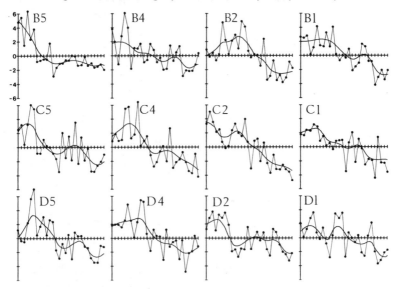

Figure 7 Graphs of the first principal components of annual fluctuations in abundance based on data for all the zooplankton taxa in each of the areas shown in the key chart in figure 4. The x-axis is in standard deviation units and each interval on the y-axis represents a year, the period covered is 1948 to 1975.

the annual fluctuations in abundance for the zooplankton in each of the 12 areas shown in figure 4. Figure 8 contains graphs of the first and second components for the whole area for each taxon. The graphs are arranged so that taxa with similar components occur together. It can be seen that long-term trends, primarily in the form of a downward linear trend or an up and down quadratic trend, account for an appreciable proportion of the year-to-year variability. There is considerable coherence between areas (figure 7) and a clear pattern of relationship between taxa (figure 8).

Thus, data from the Continuous Plankton Recorder survey can be used to monitor the open-sea plankton. The programme is being maintained and the patterns and trends are updated annually.

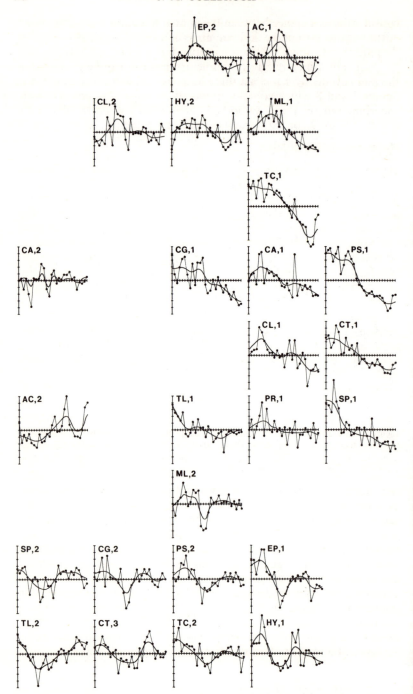

Figure 8 Graphs of principal components of annual fluctuations in abundance based on data for all the areas shown in the key chart given in figure 4. Up to two components (with one exception, they are the first and second. indicated by 1 and 2) are given for each taxon. The x-axis is in standard deviation units and each interval on the y-axis represents a year. The period covered is 1948 to 1975. The taxa are: EP, Euphausiacea; AC, *Acartia clausii*; CL, *Calanus* spp. stages V–VI; HY, Hyperiidea; ML, *Metridia lucens*; TC, Total Copepods; CA, *Candacia armata*: CG. Chaetognatha; PS, *Pseudocalanus elongatus*; CT, *Centropages typicus*; TL, *Temora longicornis*; PR, *Pleuromamma robusta*; SP, *Spiratella retroversa*.

The interpretation problem

Monitoring the plankton, with presentations expressed as descriptions of events, is not entirely without value. Obviously, however, it is desirable to explain why the observed changes have taken place. There are at least three paths that can be explored. Firstly, we can examine the data themselves; we can consider the nature of the changes that have taken place over the last three decades and we can search for patterns of relationship between species and between areas. Secondly, we can look for empirical relationships between the plankton and some of the more obvious environmental factors. And thirdly, we can study particular processes by the development of models.

Considering firstly the information available from a study of the plankton data, the changes illustrated in figures 7 and 8 suggest that persistent, large-scale factors, such as those associated with changes in the climate of the ocean, may be involved in determining fluctuations in the abundance of the plankton.

Expressions of relationship between taxa with respect to annual fluctuations in abundance have been established and compared with analogous patterns with respect to both seasonal and geographical variation. Contrary to what might be expected, it cannot be said that taxa which occur at the same time of year show similar annual fluctuations in abundance. Rather, there is evidence to suggest that taxa with similar geographical distributions show similar annual changes. Moreover, this statement is true within individual areas. Irrespective of the location of the area, whether it be oceanic or coastal, oceanic species tend to be similar to one another and neritic species tend to be similar to one another. This, together with the marked coherence between areas and taxa, suggests that, with respect to annual fluctuations in abundance on a decadal time-scale, density-independent processes play a major role in the control mechanisms.

This provides some justification for looking for, at least in the first instance, coincident and linear relationships between the annual fluctuations in the abundance of the plankton and environmental variables.

Figure 9 presents in summary form the situation with respect to two

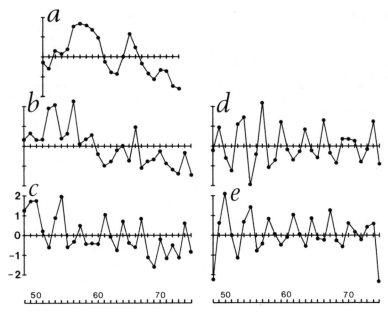

Figure 9 (*a*) A graph of the first principal component of annual variations in surface temperature and salinity at Ocean Weather Stations *India* and *Juliet* (North-east Atlantic) for 1951 to 1973. (*b*) A graph of the first principal component of the annual fluctuations in the abundance of zooplankton in area C4 (see chart in figure 4) for the period 1948 to 1975. (*c*) A graph of the annual variation in the frequency of days of westerly weather type over the United Kingdom for the period 1948 to 1975. (*d*) and (*e*) are versions of (*b*) and (*c*) respectively, filtered to emphasize 3–3½ year periodicities. All the graphs are standardized to zero mean and unit variance and the *y*-axis represents standard deviations.

environmental variables dealt with in greater detail by Colebrook (1978). Graph *a* is the first principal component of annual means of sea surface temperature and salinity at Ocean Weather Stations India and Juliet (in the North Atlantic close to the 20 °W meridian). This is an index of movement of the surface waters which, in the area of the two stations, is primarily associated with the North Atlantic Drift. Graph *c* shows the relative year-to-year variations in the number of days of westerly weather type over the United Kingdom (Lamb, 1969), which, in terms of influence on the plankton, can probably be interpreted as representing changes in a rather complex pattern of wind-driven advection. Graph *b* is the first principal component of the annual fluctuations in the abundance of 16 taxa of zooplankton in an area covering the continental shelf off the west and south coasts of Scotland between 55 ° and 60 ° North.

There are clearly similarities between the long-term trends in the three variables. In addition, the application of frequency filters to the plankton and westerly weather type variables indicates the presence of a 3 to $3\frac{1}{2}$ year periodicity in each and, as shown in graphs d and e, there is an inverse relationship between them.

Figure 10 is a histogram showing the relative proportions of various frequency peaks in the annual fluctuations in the abundance of the

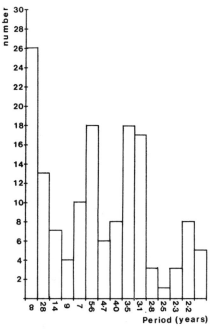

Figure 10 A histogram of the number of peaks in power spectra corresponding to various periods for the main periodic elements derived from the first and second principal components of annual fluctuations in the abundance of zooplankton for 1948 to 1975 (see text).

plankton. The first and second principal components for each of 12 areas and 18 taxa were analysed using frequency filters. The resulting variables, containing representations of the dominant frequencies, were subjected to power spectrum analyses and the numbers of peaks in the spectra relating to the various frequencies were counted.

There is an obvious peak in the low frequency band, reflecting the long-term trends in the annual fluctuations. There is also a peak at 3 to $3\frac{1}{2}$ years. Most of the examples are in phase and show a negative relationship with the corresponding element in the variation of westerly weather type.

In addition to these there are also peaks at about 2 years and at 5 to 6 years which may be due to interaction between the 2 year and 3 to $3\frac{1}{2}$

year periods. Quasi-biennial oscillations in climatic factors are well
known for the southern hemisphere (see, for example, Trenberth, 1975)
and a similar cycle has recently been detected in stratospheric temperatures over central Europe (Schwentek, 1977). Figure 11 shows a comparison between stratospheric temperatures for 1958 to 1976 and the

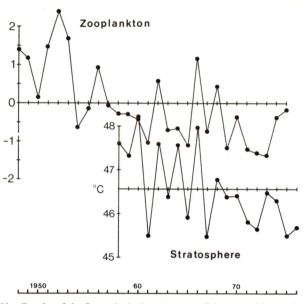

Figure 11   Graphs of the first principal component of the annual fluctuations in abundance of zooplankton in area C5 (see chart in figure 4) and annual means of stratospheric
temperature over central Europe.

first principal component of zooplankton for an area in the oceanic
Atlantic (area C5, 55–59 °N, 11–19 °W, see figure 4); the two sets of
data show corresponding biennial oscillations from 1959 to about 1971.
. These relationships between plankton and environment have been
presented here both to provide examples of the kinds of relationship
that are found and to illustrate some of the problems involved in their
interpretation.

In no case is it possible, as yet, to provide interpretations expressed
in terms of rates and processes. In the field of climatic change, many
parameters appear to be closely linked, which leads to difficulties in
differentiating between several possible factors which may influence
the plankton. Some of the relationships are extremely tenuous. It is not
reasonable, for example, to postulate a direct effect of stratospheric
temperature on the plankton. This relationship can, at best, only pro-

vide the starting point for a search for quasi-biennial phenomena and it may well prove to be purely coincidental.

At the same time, there is a growing interest in climatic change and in climate-related phenomena. There has been a marked improvement in techniques of data acquisition in recent years, particularly in the field of satellite observations, and there has been a considerable improvement in the availability of data relating both to historical series and to current observations.

These trends imply that there is ample scope for further work along these lines which could well provide the bases for more refined interpretations of the relationships between plankton and environment.

The consideration of variations in the standing stock of plankton primarily in terms of empirical, linear relationships with environmental variables obviously has its limitations. We must expect that the response of the plankton to any perturbation has a transient as well as a steady state component. As the environment is in a state of continuous change, the existence of a transient phase can complicate the relationships between concomitant values of input and response. In addition, plankton populations obviously exhibit persistence, for periods at least related to their generation times, and this imposes certain characteristics on sets of observations expressed as time-series. Given the limitations in the observational data, these problems can be solved only by the development of models of the processes involved. And. while simulation studies related specifically to the Continuous Plankton Recorder survey are being pursued, it would be premature, at this stage to report on their progress.

Acknowledgements
This work forms part of the programme of the Institute for Marine Environmental Research, which is a component of the Natural Environment Research Council; it was commissioned in part by the Ministry of Agriculture, Fisheries and Food.

References
Colebrook, J. M. (1960) Continuous plankton records: methods of analysis 1950–1959. *Bulletins of Marine Ecology*, **5**, 51–64.
Colebrook, J. M. (1975a) The continuous plankton recorder survey: automatic data processing methods. *Bulletins of Marine Ecology*, **8**, 123–142.
Colebrook, J. M. (1975b) The continuous plankton recorder survey: computer simulation of some aspects of the design of the survey. *Bulletins of Marine Ecology*, **8**, 143–166.
Colebrook, J. M. (1978) Continuous plankton records: zooplankton and environment, North-east Atlantic and North Sea, 1948–1975. *Oceanologica Acta*, **1**, 9–23.
Glover, R. S. (1967) The continuous plankton recorder survey of the North Atlantic. *Symposium of the Zoological Society of London*, **19**, 189–210.
Hardy, A. C. (1939) Ecological investigations with the continuous plankton recorder: object, plan and methods. *Hull Bulletins of Marine Ecology*, **1**, 1–57.
Lamb, H. H. (1969) The new look of climatology. *Nature, London*, **223**, 1209–1215.

Oceanographic Laboratory, Edinburgh (1973) Continuous plankton records: a plankton
   atlas of the North Atlantic and the North Sea. *Bulletins of Marine Ecology*, **7**, 1–174.
Rae, K. M. (1952) Continuous plankton records: exploration and methods, 1946–1949.
   Bulletins of Marine Ecology, **3**, 135–155.
Taylor, C. G. (1953) Nature of variability in trawl catches. *Bulletin of the U.S. Bureau of
   Fisheries*, **54**, 145–166.
Trenberth, K. E. (1975) A quasi-biennial standing wave in the southern hemisphere and
   interrelations with sea surface temperature. *Quarterly Journal of the Royal Meteoro-
   logical Society*, **101**, 55–74.
Schwentek, H. (1977) A quasi-biennial period in stratospheric temperature over Central
   Europe. *Journal of Interdisciplinary Cycle Research*, **8**, 220–221.

The monitoring of substances in marine waters for genetic damage

JAMES M. PARRY and M. A. J. AL-MOSSAWI

Department of Genetics, University College of Swansea, Swansea, Wales

Introduction

The use of the marine environment as a sink for the disposal of chemicals both deliberately and accidentally suggests that at least a fraction of the living organisms found in the seas and oceans may be exposed to potentially mutagenic agents. Such exposure may result in changes in the genetic architecture of marine populations and if such agents enter food chains they may lead to the unwitting exposure of human populations.

A variety of screening systems have been developed which involve the use of biological indicator organisms for the detection of the possible mutagenic activity of environmental chemicals. Such systems involve the use of organisms varying from relatively simple bacteria to mammalian cells in tissue cultures (see Scott, Bridges, and Sobels, 1977). All the systems utilize the procedure of exposing cells of known genotype to test agents and the plating of treated cells upon selective medium to detect those cells which have been mutated by the agent. In a large number of cases mutagenic activity in the screening systems has been shown to be correlated with the ability of a chemical to produce tumours in either experimental animals or humans (McCann *et al.*, 1975).

The assay of the seas and oceans for the presence of mutagenic chemicals can be performed in two fundamentally different ways. One method involves chemically analysing ocean samples in order to identify constituent chemicals which may then be individually tested for mutagenic activity. This is a formidable task because of the very large number of chemicals, both natural and man-made, that are detectable and must therefore be screened for mutagenic activity. However, studies of this type have been performed on the constituents of some samples of drinking water in the USA and have revealed the presence of mutagenic chemicals (Simmon, Kauthanen, and Tarditt, 1977).

The second approach involves the collection of ocean samples, concentration of the constituents, and the exposure of the resulting concentrate to a range of mutagenic screening systems. However, this second procedure has at least two fundamental limitations, the prob-

103

lem of developing a suitable concentration procedure and, perhaps more importantly, the difficulty of detecting mutagens in the presence of toxic chemicals. For example, a mutagenic chemical might be undetectable if the assay is performed in the presence of a chemical which is highly toxic to the test organism.

We have attempted to overcome the problems of the latter procedure by using marine organisms as concentrators of potentially mutagenic chemicals. By the use of such living organisms we have also selected against the possible masking effects of toxic chemicals on the assumption that such toxicity would lead to the death of the marine species in question. We have developed an assay system based upon the extraction of tissues derived from the mussel *Mytilus edulis* and the screening of these extracts for mutagenic activity using a number of microbial indicator species.

Materials and methods

Strains

The strains of bacteria used in these studies included cultures of *Salmonella typhimurium*, auxotrophic for histidine (kindly provided by Dr Bruce Ames) and a variety of *Escherichia coli* strains (kindly provided by Dr G. Mohn, Dr M. Green, and Dr D. Tweats or constructed in our laboratory) and have all been described elsewhere (Parry, Tweats, and Al-Mossawi, 1976). The strains used were capable of detecting both frameshift and base-substitution mutagens.

The yeast strain *JD1* was auxotrophic for both histidine and tryptophan and produces prototrophic colonies by the process of mitotic gene conversion, a process of genetic change that responds to treatment by mutagens and carcinogens in an essentially non-specific manner. The use of this yeast strain has been described in detail elsewhere (Davies, Evans, and Parry, 1975).

Preparation of mussel extracts

Samples of the mussel, *Mytilus edulis* were collected from a variety of sites, washed with running water, and either extracted immediately or kept frozen at $-20\,°C$ until required. (There was no evidence in our work of any reduction in genetic activity during storage.)

For the preparation of extracts of whole mussel tissue, 50 shelled mussels were extracted in 100 cm³ of 95 per cent ethanol, after disintegration of the tissue in a Waring blender or Atomix. The resulting tissue homogenate was left to stand overnight at 4 °C, resuspended, and centrifuged at 5000g for 15 minutes. The resulting supernatant was sterilized by passing through a membrane filter and stored at 4 °C for use within one week and at $-20\,°C$ for long-term use. All the samples obtained were tested for the presence of radioactive material. None of the samples described here showed levels of radioactivity above the background.

Detection of genetic activity

Fluctuation tests were carried out, with minor modifications, as described by Green, Muriel, and Bridges (1976). Overnight cultures of *E. coli* tester strains were prepared in supplemented Davis-Mingioli minimal medium. After being washed twice in saline, the cells were resuspended in saline at a concentration of 5.0×10^7 cells per cm^3. To 100 cm^3 of Davis-Mingioli basal salts was added 0.7 cm^3 of 40 per cent glucose, 0.1 cm^3 of tryptophan solution (200 $\mu g/cm^3$) for fluctuation tests involving tryptophan auxotrophs only, or 0.1 cm^3 of lysine solution (200 $\mu g/cm^3$) for lysine auxotrophs only, 0.1 cm^3 of washed cells and, in the case of the test treatments, either methyl methane sulphonate to a final concentration of 1 $\mu g/cm^3$ as a positive control, or 0.1 cm^3 of alcoholic mussel extract. In the case of experiments involving the use of multiple auxotrophs the supplements required by the non-selected markers were added in excess. Control experiments involving mussel extracts, also contained 0.1 cm^3 of 95 per cent ethanol as a negative control. Each treatment was dispensed in 2 cm^3 samples into 50 test-tubes or in some cases 1 cm^3 samples into 100 test-tubes. Once the auxotrophic bacteria had exhausted the small amount of supplement present, only prototrophic revertants continued to grow. From two days onwards, tubes in which mutation had occurred became turbid, while other tubes remained relatively clear. The number of turbid tubes was routinely scored after three days. The significance of the response of each set of 50 tubes to the presence of mussel extract was determined by the use of Chi-square analysis as described by Green *et al.* (1976). The arginine-56 mutation used in some experiments is leaky and as a result some residual growth was observed even in the absence of arginine and, therefore, trace amounts of arginine were not added when this marker was used.

In a number of experiments the fluctuation test was modified according to the procedure developed by Dr David Gatehouse (personal communication). In this modification the samples were added to 96-well microtitre plates (Sterilin Ltd) in the form of 0.2 cm^3 aliquots per well. All other steps in the procedure were as described above.

In the yeast assays, we used stationary phase cultures of the strain *JD1* which were suspended in saline at a concentration of 10^7 cells/cm^3. Samples of the mussel extracts (up to 4 per cent) were added to yeast minimal medium at $45\,°C$ together with yeast cells at a concentration

of 35 × 10³ cells/cm³ for the detection of histidine-independent prototrophs and 35 × 10² cells/cm³ for the detection of tryptophan-
independent prototrophs and poured into 9 cm Petri dishes. The culture
medium was supplemented with 20 μg/cm³ tryptophan and 0.1 μg/cm³
histidine for the detection of histidine-independent prototrophs and
with 20 μg/cm³ histidine and 0.1 μg/cm³ tryptophan for the detection
of tryptophan-independent prototrophs. These supplements enable the
auxotrophic cells to undergo three cell divisions in the presence or
absence of mussel extracts. All plates were grown in the dark at 28 °C
and scored after nine days of incubation.

In all experiments involving the yeast cultures, we used at least five
replicate plates per treatment and all experiments have been performed
at least three times. For each positive sample, appropriate dose-
response curves have been produced but in this paper we reproduce a
set of typical results for each geographic and tissue sample.

Results

Geographic variation in the presence of genetically active chemicals

Alcoholic extracts of the whole tissue of mussels collected from a
variety of geographical locations were screened for the presence of
genetically active chemicals. The techniques used were bacterial fluctuation tests which measure induced mutation and by the measurement of
induced mitotic gene conversion in the yeast *Saccharomyces cerevisiae*.
Both systems involve the measurement of the frequencies of prototrophic
cells produced in auxotrophic cultures.

Some of the results involving the measurement of induced mitotic
gene conversion in the yeast strain *JD1* are shown in table 1. The results
presented indicate the presence of genetically active material in the
samples collected from Plymouth, Caswell Bay, and Mumbles, whereas
no such activity was detectable in samples collected from Anglesey.

Further confirmation of the site variation in genetic activity was
obtained by examining alcoholic extracts of mussels collected on the
same day at a series of sites westward from Mumbles. These samples
were screened for genetic activity by the measurement of mitotic gene
conversion using the yeast strain *JD1*. The results obtained from these
assays are shown in table 2 and they demonstrate that genetic activity
decreased in the samples collected at increasing distances from the
Mumbles site. The decreased genetic activity correlates with reduced
visual pollution and increased distance from industrial areas.

Tests for the presence of nutrients in the mussel extracts

The majority of the tests to detect genetic activity in mussel tissues are

Table 1 The induction of mitotic gene conversion in yeast in the presence of a variety of mussel extracts

Treatment	Mean no. of his$^+$ prototrophs per plate	his$^+$ prototrophs per 10^6 survivors ±s.e.	Mean no. of trp$^+$ prototrophs per plate	trp$^+$ prototrophs per 10^5 survivors ±s.e.	Percentage viability
Control 1	8.6	26.8 ± 4.6	7.4	23.1 ± 4.3	91.4
Control 2	7.8	32.5 ± 5.8	5.2	21.7 ± 4.8	68.6
4% Alcohol control 1 ⎱ negative controls	6.5	25.6 ± 5.0	9.4	37.0 ± 6.0	73.1
4% Alcohol control 2 ⎰	4.2	11.8 ± 3.0	7.6	21.4 ± 3.9	101.4
Plymouth extract 4%	202.6	493.6 ± 21.3	154.8	463.5 ± 18.6	95.4
Mumbles extract 4%	71.2	323.6 ± 19.2	85.3	387.7 ± 20.6	62.9
Caswell Bay 4% extract	69.0	276.0 ± 16.6	69.2	276.8 ± 16.6	71.4
Anglesey 4% extract	4.7	18.3 ± 4.2	7.6	29.6 ± 5.4	73.4
Ethyl methane sulphonate (1 µg/cm³) —positive control	284.7	900.9 ± 26.7	346.2	1,095 ± 29.4	90.3

Table 2 The induction of mitotic gene conversion in yeast in the presence of mussel extracts collected in the Gower area of South Wales

Treatment	his⁺ prototrophs per 10^6 survivors ± s.e.	trp⁺ prototrophs per 10^5 survivors ± s.e.	Percentage cell viability
Control 1 4% ethanol	19.7 ± 1.9	7.1 ± 1.2	84.2
Control 2 4% ethanol	24.2 ± 2.2	8.9 ± 1.3	73.3
Mumbles 4%	621.4 ± 11.1	247.5 ± 7.0	65.4
Caswell Bay 4%	492.9 ± 7.0	132.0 ± 3.6	78.5
Oxwich 4%	34.2 ± 1.8	15.1 ± 1.2	82.2
Port Eynon 4%	29.0 ± 1.7	9.2 ± 0.9	74.0
Rhossilli 4%	31.3 ± 2.5	6.4 ± 1.1	68.6

based upon the measurement of the production of prototrophic colonies in auxotrophic microbial cultures. Assays of this type may lead to inconclusive results if the samples contain significant amounts of the specific nutrient deleted from the selective media. In such cases, additional growth of the test culture would lead to a spurious positive result Thus it was necessary for us to ensure that all the samples did not contain significant amounts of nutrients.

In the case of the bacterial fluctuation tests, superficial estimates of the presence of nutrients can be made by visual and colorimetric examination of the clear tubes, i.e. those that show only background growth due to the presence of the original auxotrophic cells. In those cases where significant amounts of nutrients are present in the mussel extracts, increased background growth in the tubes was readily observable. In those cases where free nutrients were present, the samples were rejected for assay or we utilized a mutagenicity assay not involving the use of the contaminating nutrient.

Those samples which showed positive genetic activity were further investigated for the presence of free nutrients by making viable counts of the numbers of bacterial cells present in both the clear and turbid tubes in the fluctuation tests. A typical example of the results of such an assay is shown in table 3, which shows the viable cell counts made

Table 3 Viable cell counts of turbid and clear tubes from a fluctuation test involving the use of *Escherichia coli* culture uvr A trp (R46) with and without the addition of mussel tissue extract from the Mumbles area

Treatment	Mean value turbid tubes (cells/cm³)	Mean value clear tubes (cells/cm³)	Median value clear tubes (cells/cm³)
Control	5.9×10^8	5.3×10^6	1.27×10^7
Control + 0.1 m ethanol	6.0×10^8	5.0×10^6	1.95×10^7
0.1 cm³ Mussel extract	5.3×10^8	5.6×10^6	9.0×10^6

from individual fluctuation test-tubes using the *Escherichia coli* culture uvrAtrp(R46), in both the presence and absence of Mumbles mussel extract. The results in the table demonstrate that the median and mean numbers of viable bacterial cells per clear tube, that is in those tubes that do not contain a significant number of trp$^+$ revertants, was unaffected by the presence of mussel extracts. Therefore, we can conclude that the particular sample of Mumbles mussel extract did not enhance the growth of trp$^-$ cells when the added tryptophan supply had become exhausted. We can further conclude that the observed increase in the frequency of turbid tubes in the bacterial fluctuation test was not due to increased cell division in the presence of free nutrients, which would have allowed the production of an increased frequency of spontaneous mutants.

The possible nutrient effects of the mussel extracts upon auxotrophic yeast cultures were determined in two different ways:

1. After the incubation period, the plates of selective agar were scored for the frequency of prototrophic colonies and then discs of agar of 9 mm diameter were removed from areas of the plates showing background growth, but no large prototrophic colonies. Agar discs sampled from 10 plates per treatment were added to 20 cm^3 of sterile saline and the yeast cells contained in the discs were suspended by sonication. After appropriate dilutions the saline suspensions were plated upon yeast-complete agar and the colonies produced were counted after five days' incubation at 28 °C. Comparisons of the viable cells contained in the background agar of both the treated and untreated cultures could then be used to determine if any further growth of auxotrophic yeast cells had taken place in the solid medium in the presence of the mussel extract. The results of such an assay using the mussel extract derived from the Plymouth site are shown in table 4 and demonstrate that there are no significant differences in the growth of auxotrophic cells between the plates with and without mussel extract.

Table 4   Assay of the background growth of yeast cells from the plates of selective agar with and without the presence of added mussel extract (Plymouth)

Treatment	Background growth, mean of discs removed from selective plates used for the detection of his$^+$ prototrophs (cells/cm^3)	Background growth, mean of discs removed from selective plates used for the detection of trp$^+$ prototrophs (cells/cm^3)
Control	$2.9 \pm 0.5 \times 10^5$	$4.1 \pm 0.8 \times 10^4$
Control + 1 % ethanol	$3.2 \pm 1.1 \times 10^5$	$3.7 \pm 1.2 \times 10^4$
1 % Mussel extract	$2.4 \pm 0.7 \times 10^5$	$3.5 \pm 1.1 \times 10^4$
2 % Mussel extract	$1.9 \pm 0.8 \times 10^5$	$4.3 \pm 0.8 \times 10^4$
4 % Mussel extract	$1.7 \pm 0.4 \times 10^5$	$4.0 \pm 1.3 \times 10^4$

2. Samples of the mussel extracts were added to liquid minimal medium, containing either excess histidine or excess tryptophan together with 10^5 yeast cells/cm³ of strain *JD1* auxotrophic for both histidine and tryptophan, and the cultures were aerated for periods of up to 48 hours. During this period cell growth was determined by counting cell numbers using a haemocytometer. Under these conditions, samples which contain nutrients show increases in cell numbers during the first six hours of incubation. In those samples lacking nutrients, no increases in cell numbers were observed until after a period of approximately 24 hours when the growth of pre-existing prototrophic cells becomes significant. The results of a typical experiment of this type are shown in figure 1.

None of the mussel samples reported here to be mutagenic were demonstrated to contain significant amounts of the specific nutrient required by the test microbes. However, we have detected such nutrients in many samples and in particular we have found that extracts of fish tissue which have also been studied in our laboratory have been shown to contain significant amounts of free amino acid particularly of histidine.

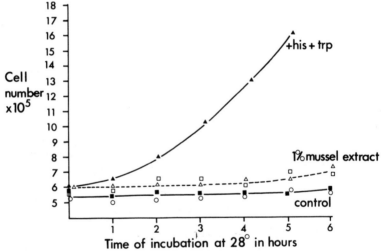

Figure 1 The effects of the addition of mussel extract upon the growth of yeast cells of strain JD1 in yeast minimal medium: ▲—medium supplemented with 20 µg/cm³ of histidine and tryptophan; ○—medium supplemented with 20 µg/cm³ histidine only; ■—medium supplemented with 20 µg/cm³ histidine and 1 per cent ethanol; □—medium supplemented with 20 µg/cm³ histidine and 1 per cent mussel extract; △—medium supplemented with 20 µg/cm³ tryptophan and 1 per cent mussel extract.

The assay of specific mussel tissues

The experiments described above demonstrate the presence of genetically active chemicals in alcoholic extracts derived from all the tissues of the organism. We have also determined whether this activity is distributed throughout all the tissues of the organism or concentrated within specific tissues.

Initially we separated the body tissues into (a) the hepatopancreas and (b) all the remaining tissues. Ten grams of each of sample (a) and sample (b) were then extracted in 50 cm³ of 95 per cent ethanol. These extracts of hepatopancreas and the remaining tissues were screened for genetic activity using a range of bacterial strains in a series of fluctuation tests. The results of these fluctuation tests are shown in table 5 and they

Table 5 Effects of mussel extracts from a variety of different tissues upon induced mutation, measured in bacterial fluctuation tests using *Escherichia coli*

Strain	Locus	Treatment	No. of tubes sampled	No. of tubes positive		Significance (probability)
				Control	Treated	
341/113	arg 56	MMS	50	24	41	< 0.001
		Hepatopancreas	50	24	34	NS
		Remaining tissue	50	24	41	< 0.001
341/113	arg 56	MMS	50	28	48	< 0.001
(R46)		Hepatopancreas	50	35	41	NS
		Remaining tissue	50	35	45	< 0.01
uvr A	trp	MMS	100	54	87	< 0.001
(R46)		Hepatopancreas	100	48	59	NS
		Remaining tissue	10	48	70	< 0.005
343/113	arg 56	MMS	50	21	39	< 0.001
		Mantle	50	24	34	> 0.05
		Foot	50	24	34	NS
343/113	lys 60	MMS	50	28	40	< 0.02
(R46)		Mantle	50	28	40	< 0.02
		Foot	50	28	35	NS
uvr A	trp	Mantle	50	15	28	< 0.02
(R46)		Foot	50	15	18	NS
		Reproductive system	50	15	18	NS

MMS = positive control methyl methane sulphonate
NS = not significant

demonstrate that for the mussel samples tested in this way genetic activity was not detected in the extract of the hepatopancreas but was to be found in the remaining tissue.

In order to characterize further the tissue containing the genetically active material, mussels collected from the Mumbles site were dissected and 10 g samples were isolated of muscle, mantle, foot, and the reproductive system, which were then each extracted in 50 cm³ of 95 per cent ethanol. Samples of each of the tissue extracts were assayed for the presence of genetically active chemicals by the measurement of induced mutation in bacteria in a series of fluctuation tests and by the measurement of induced mitotic gene conversion in yeast.

The results of the assays of the various mussel tissues for genetic activity are shown in table 6. Both the mutation tests with bacteria and those of induced mitotic gene conversion in yeast demonstrate that the major site of genetically active chemicals in mussels collected from the Mumbles site was the mantle tissue.

Table 6 Effects of mussel extracts from a variety of different tissues upon the induction of mitotic gene conversion in yeast strain JDI

Treatment	Viability	trp+ prototrophs per 10^5 survivors ±s.e.	his+ prototrophs per 10^6 survivors ±s.e.
Control	100.0	19.1 ± 1.8	36.7 ± 2.3
Negative control 2% ethanol	101.3	22.4 ± 1.7	32.4 ± 1.9
Muscle 1%	97.3	16.6 ± 1.4	37.5 ± 2.3
2%	94.4	29.3 ± 1.8	41.0 ± 2.9
Mantle 1%	101.9	94.5 ± 3.9	170.3 ± 11.6
2%	93.2	135.7 ± 8.2	589.6 ± 23.4
Hepatopancreas 2%	85.3	27.9 ± 1.5	41.6 ± 2.7
Foot 2%	92.4	17.6 ± 0.9	39.4 ± 2.1
Reproductive system 2%	76.8	14.3 ± 2.3	20.7 ± 1.9

Discussion

The results presented here demonstrate the practical value of microbial assay systems for the detection of genetically active chemicals in the tissues of the mussel *Mytilus edulis*.

We have been able to demonstrate that from selected sites around the coast of the United Kingdom mussels may be collected which contain within their mantle tissue chemicals, which are extractable in ethanol, and that these are capable of inducing mutation in bacteria and mitotic gene conversion in yeast. The presence of nutrients such as free amino acids in the tissues of many marine species prevents the uncritical use of the standard microbial mutagenicity assay systems with extracts of all marine organisms. However, with appropriate technical modifications we have been able in our laboratory to utilize the techniques with a diverse range of marine organisms such as plankton, oysters, crabs, and a number of fish species.

The techniques described here demonstrate only the presence of genetically active materials and, although they may provide indirect information such as the mode of action of the mutagens concerned, they do not provide direct information on the chemical nature of the contaminating agents. However, a variety of chemical techniques is available for the separation of the constituents of the mussel extracts into samples which differ in their solubility. Individual fractions, in different solvents, may then be tested separately for possible genetic activity. Separation of constituents in this way considerably simplifies the task of the analysis, which can then be used to determine the chemical constitution of any active fractions. Once information becomes available on the individual constituents of the active fractions the microbial assay systems may then be used to determine the specific genetic activity of each chemical.

The individual mussel samples collected from the different sites probably contain a diverse collection of chemicals and it is unlikely that the activity we detect derives from either a single or a group of chemicals. It is likely that each of the sites represents a different pool of genetically active chemicals all of which will require characterization.

Although our results demonstrate the presence in mussels of chemicals which cause mutation in microbes, they provide no direct information on the potential of these chemicals to produce similar changes in the cells of higher organisms. However, our techniques do provide useful information on the chemicals in the marine environment which have the potential to produce genetic damage in those organisms exposed directly or at some stage in marine food chains. After the identification of the active chemicals by the use of the techniques we have described, the evaluation of their potential hazards will require the use of a range of test systems of more direct relevance to the organisms of concern.

Acknowledgement
We would like to dedicate this paper to Mr Joe Motley whose many discussions with the senior author initiated our interest in environmental mutagenesis.

References
Davies, P. J., Evans, W. E., and Parry, J. M. (1975) Mitotic recombination induced by chemical and physical agents in the yeast *Saccharomyces cerevisiae. Mutation Research,* **29**, 301–314.
Green, M. H. L., Muriel, W. J., and Bridges, B. A. (1976) Use of a simplified fluctuation test to detect low levels of mutagens. *Mutation Research,* **38**, 33–42.
McCann, J., Choi, E., Yamasaki, E., and Ames, B. N. (1975) Detection of carcinogens as mutagens in the Salmonella/microsome test: assay of 300 chemicals. *Proceedings of the National Academy of Sciences,* **72**, 5135–5147.
Parry, J. M., Tweats, D. J., and Al-Mossawi, A. J. (1976) Monitoring the marine environment for mutagens. *Nature,* **264**, 538–540.
Scott, D., Bridges, B. A., and Sobels, F. H. (1977) *Progress in Genetic Toxicology.* Amsterdam: Elsevier/North Holland.
Simmon, V. F., Kaukanen, K., and Tardift, R. G. (1977) Mutagenic activity of chemicals identified in drinking water. In *Progress in Genetic Toxicology,* ed. Scott, D., Bridges, B. A., and Sobels, F. H. pp. 249–258. Amsterdam: Elsevier/North Holland.

Monitoring whale and seal populations

RICHARD M. LAWS

*British Antarctic Survey and Sea Mammal Research Unit,
Natural Environment Research Council,
Madingley Road, Cambridge, England*

Introduction

All ecosystems are in process of change and the changes may be long
or short term and cyclic or unidirectional. Long-lived animals such as
whales and seals—with life spans in the range 20 to 100 years—tend
to require long-term studies to detect changes, although some short cuts
may be possible. Data are often collected for their own sake but it is
desirable to have a central hypothesis to test; in the case of marine
mammals the issue in simple terms is usually whether the population
is decreasing, stable, or increasing. For many marine mammals this is
surprisingly difficult to establish. In the report of a recent meeting on
small cetaceans (Mitchell, 1975) 66 species were considered; there was
no quantitative estimate of population size or trends in abundance for
all except eight and even these were inadequate. For those species such
as the larger whales that are or have been the basis of an industry there
are usually more data available. Management of marine mammals still
tends to be based on the assumption that the ecosystem is stable, but
Strange (1972) has shown that large-scale changes can occur in the
absence of direct human intervention and Stirling, Archibald, and De
Master (1977) have shown that these changes can be rapid. Quotas
calculated for a stable healthy population could be devastating to a
reduced population with lowered productivity.

In addition there is a growing awareness that technological developments are bringing industry to remote areas. Large-scale oil exploration
and exploitation in inshore and offshore areas is under way, and large
industrial plants producing noxious effluents are being established in
areas where they could affect marine mammal populations. Even the
Antarctic is not immune. So far there is no evidence that marine
mammals are seriously affected by oil-spills but the less obvious effects
of disturbance, such as noise, may be important.

The oceans are vast (361 million km^2). While some marine mammals
are localized in distribution as year-round residents or seasonally as
breeding concentrations or along migration routes, others are very
widely dispersed. In general, most of the seals are in the first group and

most of the cetaceans in the latter, but there are exceptions. However, the degree of concentration has important consequences for the carrying out of monitoring operations.

A stock that concentrates at some time during the annual cycle can be counted relatively easily and cheaply, and with greater precision, either on the ground or from the air. If, like all seals, it is hauled out on land, sandbanks, or ice for breeding, it is possible to count different components of the population by sex and broad age-classes. Similarly, a whale population that follows well-defined narrow migration routes or concentrates in localized areas for calving can be counted from boats, land, or air.

The more dispersed stocks present difficulties. In the case of the whales, if the sea is rough, visibility poor, or a strong wind dispersing the blow, the chance of sightings is severely reduced. Also, because they are dispersed, sightings tend to be infrequent; observer fatigue is a problem.

Rarely is a whale or seal population evenly distributed over its range. There may be segregation by age, size, sex, and reproductive state. Thus background knowledge of the general biology and behavioural ecology is essential for interpretation of counts. Adjustments are necessary, even in the simplest case of a colony breeding seal, for the immature segment of the population and for the time of the counts. Even in species in which births are closely synchronized the count date is critical and counts must be corrected by reference to the curve of births over time.

As the difficulty of counting increases, so the variances of the resulting estimates of population size grow and, if too large, then trends in population size are difficult to identify because successive estimates are not significantly different. The time scale over which changes become apparent may be unacceptably drawn out for a useful monitoring operation.

For this reason indices are often more useful—it is information on relative abundance, status, and rate of change that is sought. Indices can relate to population abundance, or density, in terms either of the natural population (e.g. sightings per km^2) or often the harvest taken from a stock (catch per unit of effort, CPUE).

Indices can also relate to the quality or structure of the population— by direct observations or from sampling the population. In many respects the latter gives the most complete information but, for samples of meaningful size, usually depends on a commercially based industry. However, scientific sampling can be undertaken. For some populations sampling that does not involve killing may be possible, involving time-series observations of known individuals for growth rate, age at first maturity, and frequency of pupping or calving. If this is possible then a direct count is usually also possible, but the additional dynamic information is valuable.

Finally, recent developments have drawn attention to the advantages of retrospective examination to show trends. The age at sexual maturity can be back-calculated from ear plug or tooth structure so the trends in mean age at sexual maturity, or mean growth rates, can be determined for age-classes born decades earlier.

Abundance estimated from counts

For a number of species direct counts or estimates have been made, but in few cases have population sizes been monitored over a sufficiently long period to provide evidence of trends in abundance. Some examples are given in tables 1 and 2 of stocks not subject to harvesting, whose abundances can be documented over periods of up to 80 years. In most cases this is retrospective monitoring, because the concept is a recent development. Some, however, have been intentionally monitored for decades and a proper management plan provides for adequate monitoring and feedback. Tables 1 and 2 give estimates based on counts of the net instantaneous rates of increase or decrease (r) for 25 populations. These show a wide range of r, from -0.091 for the Falkland Islands sea lion stock to $+0.155$ for the Antarctic fur seal; and from $+0.020$ for the Antarctic blue whale (best estimate) to $+0.146$ for a southern right whale stock.

Seals on land

The northern fur seal *Callorhinus ursinus* still provides the finest example of rational exploitation of any wild stock of animals (Bertram, 1950). Following its reduction by uncontrolled sealing in the nineteenth century it was relatively easy to count classes of animals on land, and up to 1946 an annual 'computation' of the herd on the Pribilof Islands was made, that gave a spurious impression of accuracy. It was based mainly upon harem bull counts and, up to 1922, on pup counts. (The data in table 1 come from the early years of increase.) When the herd became so large that it could not be counted directly other methods had to be used, and these are still evolving.

Since 1947 pups have been tagged and recovered in the kill at ages 2–5 years. to estimate numbers alive at tagging. Tag-recapture is also used to estimate numbers of yearling and older animals but large corrections are necessary particularly for tag loss. Harem bulls are counted in June to August, but mature females cannot be counted directly because of their high density on the beaches; their numbers are estimated by dividing the estimated number of pups born, by the average

Table 1 Estimates from counts of natural increases or decreases in 18 seal populations

Species	Area		r	Author of data or estimate
Grey seal *Halichoerus grypus*	Britain	1956–1976	0.061–0.070	Summers (in press)
	Sable I.	1962–1975	0.110	Summers (in press)
N. Elephant seal *Mirounga angustirostris*	California	1890–1958	0.083	Bartholomew and Hubbs (1960)
		1890–1965	0.079	Johnson (1976)
		1890–1969	0.082	Johnson (1976)
	Ano Nuevo I.	1961–1968	0.320	Johnson (1976)
	San Miguel I.	1958–1976	0.200	Johnson (1976)
S. Elephant seal *M. leonina*	Marion I.	1951–1975	−0.051	Condy (1977)
Weddell seal *Leptonychotes weddelli*	Signy I.	1947–1967	0.000	Smith and Burton (1970)
	McMurdo Sound	1963–1974	0.018	Siniff *et al.* (1977)
Southern fur seal *Arctocephalus australis*	Uruguay	1960–1972	0.056	Brownell (1973)
	Falkland Is.	1951–1965	0.000	Strange (1972)
Sub-Antarctic fur seal *A. tropicalis*	Marion I.	1955–1974	0.105	Condy (in press)
	Gough I.	1955–1975	0.100	Condy (in press)
Antarctic fur seal *A. gazella*	South Georgia	1958–1975	0.155	Payne (1977)
	South Orkneys	1956–1971	0.084	Laws (1974)
Northern fur seal *Callorhinus ursinus*	N. Pacific	1912–1924	0.082	Lander and Kajimura (1976)
		1958–1964	0.056–0.063	Lander and Kajimura (1976)
	San Miguel I.	1973–1975	0.511[1]	Johnson (1976)
Southern sea lion *Otaria byronia*	Falkland Is.	1937–1965	−0.091[2]	Strange (1972)
Northern sea lion *Eumetopias jubata*	Aleutian Is.	1957–1975	−0.035[2]	Anon (1977b)

Note: r is the net instantaneous rate of increase in the equation $N_t = N_0 e^{rt}$. It applies to total population or subset.

[1] Includes immigration; [2] estimate of *r* based on only two points.

pregnancy rate estimated from pelagic sampling. Lander and Kajimura (1976) discuss the problems and give tabulations of pups born and harem bull numbers for the three main herds from 1911 to 1975. The annual kills over the period 1871–1975 are given and a graph illustrates the trends in the number of pups born at the Pribilof Islands and the male kills made $3\frac{1}{2}$ years later.

In recent years there has been a recolonization of parts of the former range of the species and Johnson (1976) describes the very rapid increases at a new colony on San Miguel Island, California. The numbers of pups born were 261 in 1973, 521 in 1974, and 725 in 1975. This gives an impossibly high value for r of 0.511. However, some of the mature females observed were tagged as pups on the Pribilof Islands and Commander Islands and it seems clear that the rate of increase was inflated by immigration, perhaps of young mature females, as Payne (1977) observed at South Georgia for *Arctocephalus gazella*.

These southern fur seals have also been increasing under protection, following near extinction by sealers in the nineteenth century. Payne (1977) has described a continuing study at South Georgia where *A. gazella* has shown an exponential annual increase in pup numbers of 16.8 per cent from 1957 to 1975; this represents a doubling time of 4.7 years. It is much higher than other recorded seal population increases and Laws (1977a, 1977b) suggested that it was influenced by super-abundance of its main food, krill, during the breeding season, following the reduction of the baleen whale stocks around South Georgia. Over 90 per cent of the births occur within a week and timing of the counts is important; corrections are necessary to allow for density-dependent variation in counting efficiency and a mark-recapture method is used for this. Monitoring of the density-dependence of pup survival and pregnancy rates continues and an intensive study of pup mortality in relation to density in expanding colonies has begun. Condy (in press) presents less complete data for increases of 10.5 per cent annually in *A. tropicalis* populations from 1951 to 1975 (table 1).

Another classic example is the northern elephant seal *Mirounga angustirostris*, similarly reduced to near-extinction by 1890. Bartholo-mew and Hubbs (1960) gave a population history based on counts and numbers killed from 1880 to 1958, and this study is continued by Johnson (1976). The stock appears to have been reduced to 20–100 animals on Guadelupe Island and underwent an exponential increase to 13 000 in 1958, 17 000 in 1965, and to an estimated 30 000 in 1969 (table 1). Again, as in the San Miguel fur seals, excessively high local increases were due to movement between local populations. At San Miguel Island there were annual increases in pups born of 22.5 per cent and on Ano Nuevo Island 38 per cent. The best data now come from the latter island, the site of intensive studies since 1961.

The British grey seal *Halichoerus grypus* populations are among the best documented of all seal populations. There is a large and growing literature and pup production has been monitored for many years by means of annual counts. Summers (in press) has summarized methods and results obtained. The three major stocks (Farne Islands, Outer Hebrides, and Orkney) have shown exponential increases of about 6–7 per cent annually for undisturbed populations (figure 1). Pup hunting, begun in 1960 for fishery protection reasons, has begun to affect

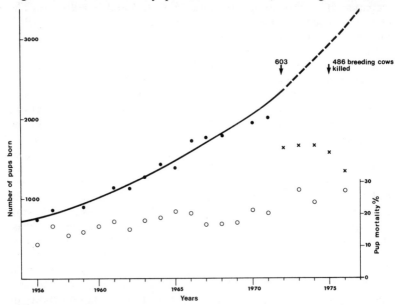

Figure 1 Increase in number of grey seal *Halichoerus grypus* pups born at the Farne Islands, 1956–1971 (black circles) and 1972–1976 after management culling (crosses). Total annual pup mortality is also shown (white circles).

recruitment to the Orkney stock and the pup production on the Farne Islands has been reduced to about 70 per cent of its 1971 level by killing adults.

Vaughan (1971) describes techniques for aerial photography of grey seals and common seals *Phoca vitulina*, using black and white and infra-red film. Around the British coasts the white-coated pups contrast with the dark ground, but where they pup on ice or snow an ultra-violet technique could be used. In the grey seal adult males, females and pups can be readily distinguished without disturbance; air and ground control counts are closely comparable. Common seals are less easy to count because only a fraction of the population is visible at any one time. For example Vaughan (1971) obtained a maximum count for the

Wash of 1700, where he knew from mark-recapture studies that the population actually numbered 4–5000.

Recently aerial surveys have been used to estimate sea lion populations. Kenyon and Rice (1961) describe the results of aerial surveys of the northern sea lion *Eumetopias jubata*. Aerial surveys in 1975–1977, compared with counts in 1957, 1960, 1965, and 1968 show a decline of 40–50 per cent in the past 20 years in the Aleutian Islands (Anon, 1977b). Strange (1972) describes aerial counts of *Otaria byronia* in the Falkland Islands in 1965 and 1966 which, when corrected, gave an estimated population of 30 000. In 1937 Hamilton (1939), from a careful ground count of pups, estimated a total population of 380 000. In a period of about 28 years there had been a 92 per cent decline in this stock, a decrease averaging 9 per cent a year.

At the same time, Strange carried out counts of the fur seal *A. australis* and compared them with a 1951 count by Laws (1953). For the same colonies visited the two counts were very close—near to 14 000—although there may have been fluctuations in the intervening years.

Seals: on ice

Stirling *et al.* (1977) describe aerial surveys of ringed seals *Phoca hispida* and bearded seals *Erignathus barbatus* on the sea-ice of the eastern Beaufort Sea, using stratified random transects from shore to 160 km offshore. Densities and population estimates indicated a decline in abundance between 1974 and 1975 of 48 and 50 per cent respectively, which could not be attributed to hunting. Lavigne, Oritsland, and Falconer (1977) described and evaluated operational remote sensing techniques and concluded that visual surveys were less effective than aerial survey techniques, especially photographic surveys which extend the human observations and provide a permanent record of the subject under study. Observation can be in the visual part of the spectrum and beyond, using ultra-violet and infra-red (false colour) wavelengths. They discuss in detail the use of photographic surveys to improve estimates of seal populations, especially the pack-ice breeding harp seal *Pagophilus groenlandicus*. It is easy to get photographs of the species, but difficult to interpret them, because male and female numbers vary with time of day, it is difficult to distinguish immatures, and the pups are difficult to see. For a brief period when pupping is completed, virtually all the young of the year are on the ice and represent the one factor that remains constant long enough for an assessment to be made. The problem was to improve the contrast of the white pup against the snow. Ultra-violet photography results in a black pup because the seal coat absorbs much of the ultra-violet radiation that is reflected by the snow. This is a very promising method and has applications to other animals. It is currently being used to improve estimates of north-western Atlantic harp seal pup production. Sergeant (1976) summarizes the history of the harp seal populations. Although harp seal management is still a controversial subject, estimates made by a variety of methods of stock assessment show that the stock size has declined substantially since 1950. (The estimates of pup production are summarized by Reeves (1977) and presented in figure 2.) Whether the predictions based on a management model that a total allowable catch of 170 000, including 136 000 pups, will allow the population to increase to MSY level (Lett and Benjaminsen, 1977) are valid remains to be seen.

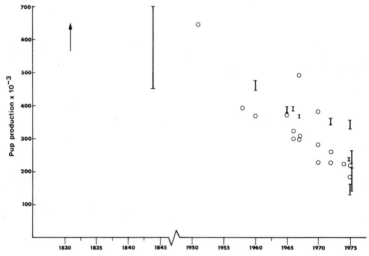

Figure 2 Summary of estimates of north-western Atlantic harp seal *Pagophilus groenlandicus* pup production. Bars indicate range estimates. (Data tabulated in Reeves, 1977.)

Whale sightings

Some whale species concentrate seasonally near shore, either for breeding or in the course of migrations along narrow pathways. They include the grey *Eschrichtius robustus*, right *Eubalaena glacialis*, bowhead *Balaena mysticetus*, humpback *Megaptera novaeangliae*, and beluga *Delphinapterus leucas*. Under these circumstances they can be counted from coastal headlands, from boats, or from the air.

Best known are the grey whale counts from the coast of California. Rice (1961) and Rice and Wolman (1971) described the census conducted from Point Loma during the migration to and from the calving grounds in Mexico. Similar counts have also been made from near Monterey since 1967/68 (Anon, 1977a). Using binoculars, counts are made from a high observation point between sunrise and sunset. The uncorrected counts give evidence of trends in abundance and were used to calculate the *r* values in table 2. The population appears to be increasing, but at a reducing rate. Rice (1961) describes how the counts can be corrected for whales missed due to foggy days, nights, and being far offshore. Rice and Wolman (1971) concluded that the population had levelled off at about 11 000. This included an assumption that speed at night was half the daytime speed, but recent observations at night using a 'starlight' scope indicate that their speed does not vary in this way (Brownell, 1977). The more recent counts suggest that population growth is still occurring but at a reduced rate.

Table 2 Estimates from sightings of natural increases in seven whale populations.

Species	Locality	Period	r	Author of data or estimate
Grey whale *Eschrichtius robustus*	California	1952–1960	0.110	Rice (1961)
		1959–1968	0.085	Rice and Wolman
		1959–1970	0.066	(1971)
		1967–1977	0.030	Anon (1977a)
Right whale *Eubalaena glacialis*	South Africa	1969–1975	0.130[1]	Best (1977)
			0.146[2]	Best (1977)
	Antarctic	recent	0.065	Anon (in press)
			0.100	Gulland (1976)
Humpback whale *Megaptera novaeangliae*	Antarctic	recent	0.065	Anon (in press)
	New Zealand	1926–1956	0.040[3]	Dawbin (1956)
Blue whale *Balaenoptera musculus*	Antarctic	recent	0.020	Gulland (1976)
	Iceland	recent	0.100	Anon (in press)

Note: See table 1 for explanation of *r*.
[1]Calves; [2]adults; [3]some harvesting of this stock.

Dawbin (1956) presented data on about 9000 sightings of humpback whales off New Zealand. For 32 years in the period 1914–1955 the numbers sighted from a whaling lookout in Tory Channel were recorded. There were but small fluctuations over 36 years in the date and duration of migrations. (Similarly, Laws (1961) showed that the mean dates of conception in Antarctic fin whales fluctuated only slightly over a 30 year period from 1925–1958.)

At Peninsula Valdez, Argentina, right whales can be closely observed and counted from clifftops or from the air and a population register has been built up from photographs of recognizable individuals (Payne, personal communication). Best (1977) describes aerial surveys of right whales in coastal waters between the latitudes of about 35° 15′ S on each coast of South Africa; he calculated the mean annual rates of increase presented in table 2. It is rather surprising that few attempts have been made to observe and count cetaceans from the air, but it seems likely that this method may be used increasingly in future because it is more cost-effective than ship surveys independent of whaling operations.

Heyland (1974), quoted by Lavigne *et al.* (1977), describes how beluga concentrate in river mouths in shallow water. Their white colour shows up clearly against the blue water and counts can be made by means of black and white photography. It is possible to identify calves and may perhaps be possible to analyse length frequencies to establish age structures as Laws (1969) and Croze (1972) have done for elephants and Watson, Graham, and Parker (1971) for crocodiles.

Gambell (1972) describes the use of a spotter aircraft for sperm whales *Physeter catodon* off Durban. More than a thousand miles of sea were covered daily during the whaling season. While catch data give biased results due to species selection in hunting, aircraft sightings are unbiased and point to random variations in sperm whale numbers from year to year. Conversely, the catch records show a fairly steady apparent increase in sperm whale density—almost certainly due to increasing interest in catching that species.

Mark returns show that fin whales *Balaenoptera physalus* off the Natal coast are related to the stocks in the Antarctic between 7° and 54° E (Area III). Best and Gambell (1968) and Best and Lurmon (1974) show that sharply declining population trends calculated for the whole Antarctic are closely followed by the pattern of availability off Durban from 1954 to 1972 (CPUE and number seen per distance flown). Similar data for sei whales *B. borealis* show a decline from the early 1960s.

Perrin, Smith, and Sakagawa (1976) concluded that in 1974 populations of spotted and spinner dolphins, *Stenella attenuata* and *S. longirostris*, in the eastern tropical Pacific numbered 3.1 to 3.5 million and 1.1 to 1.2 million respectively. The basic objective of these studies was to determine whether the populations are declining as a result of incidental catches by the tuna industry. A line transect method was used for aerial and ship survey density estimates of schools and raised by average school size estimates. They concluded that aerial survey was feasible, but population estimates were weak because of the extrapolation to large parts of the range that were unsurveyed. The work continues.

Sightings from ships also provide a direct estimate of population size, but most sightings have been provided by scouting and catching boats, so they are biased towards high density areas. The probability of sighting from boats can be very low, 0.27 for Antarctic sei whales for example. Since the extensive study reported by Mackintosh and Brown (1956) based on observations from RRS *Discovery*, the most extensive data on sightings are the surveys by Japanese whale scouting boats from the 1965/66 season onwards. Doi (1974) discussed sighting theory and developed an expression for the probability of sighting; Eberhardt (1978) describes some of the problems.Gulland (1976) reports that the best estimate gives a 2 per cent increase for blue whales *Balaenoptera musculus* (though it might well be zero or 5 per cent), 7.5 per cent for humpback, and 10.2 per cent for right whales, but with high year-to-year variation. These estimates are incorporated in table 2. A problem with whale sightings data particularly when provided by inexperienced observers—which nowadays means non-whalers—concerns species identification. It has been suggested that sightings of minke whales *Balaenoptera acutorostrata* in the Antarctic may be confused with pygmy right whales *Caperea marginata*, possibly giving an inflated population estimate for minke whale. Another possible source of error is the 'ship approach' or 'seeking' behaviour and epimeletic or 'caring' behaviour at least in Northern Hemisphere populations of bottlenose *Hyperoodon ampullatus* and minke whales. Measures of effort used for these species must allow for this behaviour (Mitchell, 1977).

Because of the possible bias and uncertainty inherent in estimates of whale numbers made from sightings at sea, the large variations they show, and the high cost of cruises independent of whaling operations, the method is not yet acceptable as an operational monitoring technique. Accounts of attempts to refine sightings theory, to improve the practice, and to analyse the results are given in papers submitted in recent years to the Scientific Committee of the International Whaling Commission and published in its Annual Reports. The FAO/ACMRR Working Party on Marine Mammals also addressed this problem in their report (Anon, 1977c). The data from sightings at sea in table 2, on right whale, humpback, and blue whale are given with appropriate reservations. For the future there is a strong need for independent sightings cruises in all areas and at all seasons, unrelated to the commercial operations. The proposal for an International Decade of Cetacean Research put forward by the Scientific Committee of the International Whaling Commission calls for a number of such cruises. It gives high priority to five stock-monitoring cruises at two-year intervals over 10 years at an estimated cost of $5 million (1975 prices). Medium priority is given to census cruises, population estimates, and the development of acoustic techniques (International Whaling Commission, 1976).

Biological indices of abundance

Even when marine mammals cannot be directly counted it is possible to estimate relative abundance from year to year by means of indices. These can be biological or related to commercial operations, and while the latter have been widely used, the former have received little attention.

Three promising methods are worth mentioning; they concern time spent searching for seal structures, frequency of vocalizations, and strandings.

Ringed seals *Phoca hispida* give birth to their young in lairs under snow on the sea-ice. From 1972 to 1975 surveys of the habitat showed a marked decrease in their density in the Amundsen Gulf area. Smith and Stirling (in press) searched 26 sites, each of 3 km^2, in 1974 and 1975. At each site they made underwater recordings of seals' vocalizations by a lowered hydrophone. Recordings were also made in 1972 and 1973 but not in exactly the same locations. In 1975 few structures (birth, haul out and male lairs, breathing holes) were found and it took longer to locate each. Search times per birth lair in 1974 were 85 and 30 minutes for inshore and offshore areas, compared with 510 and 270 minutes in 1975. In another study area, between 1973 and 1976, similar results were obtained. The number of vocalizations per minute dropped markedly each year from 1972 and 1973 to 1975, inshore from 3.33 in 1972 and 1973 to 0.06 in 1975; and offshore respectively from 2.25 to zero. The marked decline in frequency of vocalizations suggests a reduced abundance of seals below the sea-ice of a similar magnitude to that shown by other data.

Strandings of cetacea on continental coasts provide another example of an index of abundance. There has been no whaling from British coasts since 1929, but a long period of careful records of strandings may reflect abundances. Sergeant (1977) discusses the data and concludes that, given certain conditions, strandings are probably a good index of population size and this seems to apply to British stranding records. Strandings are now well monitored on the eastern seaboard of the United States and continued monitoring of strandings on European coasts should be valuable.

Mark-recapture experiments

When direct counts or the use of indices are not suitable an independent estimate of population size can be obtained by means of mark-recapture experiments. This is relatively easy for certain seals and for land mammals, but difficult for whales. The ideal mark is not yet available; it should be permanent, unique to the individual, visible from any angle, but not so conspicuous as to increase the probability of capture. Seal tags come close to this specification, but currently used whale marks are metal tubes that are fired into the body of the whale. The marked sample must be random with respect to location, size, sex, and so on and the usual assumptions about emigration, immigration, and mixing are necessary. However, as well as giving population estimates the method can, over a long term, yield additional independent information on natural mortality rates. frequency of pupping or calving, and recruitment, especially for seals.

Chapman (1974) summarizes the problems in applying the method to whales and the need to treat the results with caution. It is difficult to be sure how many whales have actually been marked, the numbers that can be marked in a season are small, recapture of current marks from large whales requires commercial whaling operations, and even when a marked whale is recaptured the probability that the mark will not be recovered is high (about 0.4). It is therefore necessary to use methods that are unaffected by these shortcomings such as the Seber-Jolly method (Seber, 1973). Results of analyses are reported in the Reports of the International Whaling Commission, but because of the large variances attached they do not provide a satisfactory monitoring operation on their own and the results have to be combined with other estimates to provide population assessments.

Colony breeding seals lend themselves particularly well to such mark-recapture experiments. For example, Siniff *et al.* (1977) have used a modified Seber-Jolly method to estimate the size of a Weddell seal *Leptonychotes weddelli* population from 1971 to 1974, a study which continues. They give estimates of the breeding population and of three components of it (parous and non-parous adult females, and adult males). The range of estimates was 1753–2401 for the total population including direct counts of pups. The use of mark-recapture in management experiments at the Pribilof Islands has already been mentioned.

Stock assessments and population models

Because of the lack of direct evidence from sightings or good biological indices of abundance it is necessary to turn to commercial indices of density and to data from the commercial catches, particularly in the case of the large whales. 'The most important single source of data for the scientific study of the whale stocks and the impact of exploitation on them, has been the records of the whaling operations themselves' (Gulland, 1976). Since the 1930s data have been recorded on the number, size, sex, and position of all whales killed, the number of catchers, and the number of days worked. If corrections are made for changes in the power and efficiency of individual catchers, the average number caught per catcher's day's work provides a good measure of abundance. Unfortunately, as species have been successively overhunted, shifts in species preferences by the whalers have complicated the picture. The basic assumption is that CPUE is proportional to population size, and there is an unknown bias due to the fact that the catches tend to be concentrated in the areas of the highest density.

CPUEs can be used to compare densities directly from year to year and many examples can be found in the reports of the IWC. They have not been widely used in seal studies because other methods of estimating abundance are usually available, but Laws (1960a) used a simple measure of catch per boat to show that elephant seal *Mirounga leonina* stocks at South Georgia were declining in the 1940s.

Mackintosh (1965) and Chapman (1974) have described the methods initially used in whale stock assessments, beginning in the 1960s, with the work of the Committee of Three Scientists. They were based on using age distributions of catches and CPUE data to compare the size of the age-classes between years, either directly from the age distribution of the catch or from applying age/length keys. Effort data are difficult to obtain for small whales but if annual catches can be divided into age-groups estimates of population size over the years could be made by the method of virtual population analysis.

With the development of more powerful computers, stock assessments have become more complex and the management policy adopted by the IWC now requires estimates of maximum sustainable yield (MSY), MSY population level, and the present population level for setting catch limits for each stock, depending on whether it is classed as an 'initial management stock', a 'sustained management stock', or a 'protection stock' (Gambell, 1976). Models provide a means of making such estimates. In order to construct a population curve which can provide an MSY estimate, information is required on natural mortality, ages at sexual maturity, and recruitment, pregnancy rates, and—for MSY by weight—growth curves; as well as the changes in these parameters with change in stock density.

Allen and Kirkwood (1977) describe two such population models for sperm whales. The first breaks up the populations into a series of blocks by sex, ages at maturity, and recruitment, and the parameters relate to unexploited and exploited populations. The second, a cohort model, divides the populations into year-class cohorts instead of age and maturity blocks. It simulates the size distributions, such that the annual catches by numbers at sizes can be removed and the results compared with the real life situation. Similar population models are used for studying other species such as minke *Balaenoptera acutorostrata*, sei, and fin whales. As knowledge increases the models can be updated and the estimates of the whale stocks are revised from year to year by the Scientific Committee of the IWC. Allen (1977) describes the model used in updating estimates of fin whale stocks for which no additional data or new population models are available.

Similar models have been developed for seal stocks of which the most recent is the harp seal model described by Lett and Benjaminsen (1977); another has been described by Capstick and Ronald (1976). They all have imperfections and the resulting estimates depend on the quality of the data inserted, especially on natural mortality rates, but they can be expected progressively to be improved and to provide in due course a sophisticated operational monitoring technique. They do not yet constitute this.

Indices of stock structure or condition

Because it is related to age and is routinely recorded in the course of whaling operations, body length should be a useful index of changes in stocks and was earlier used in this way (e.g. Laws, 1960b, 1962). There are problems however, both relating to bias in the samples and to changes in growth rates. Laws (1960b) plotted mean lengths of the catches of fin whales over a long period of years and they showed a striking segregation by longitude. It has long been known that larger and older whales tend to reach higher latitudes than smaller and younger animals. Moreover there is evidence that growth rates have increased and the mean age at sexual maturity has advanced as the density of the whale stocks has decreased. Laws (1960b, 1962) was the first to point this out, for fin whales, and Lockyer (1972, 1974) produced more evidence from ear plug studies on fin and sei whales. Also, the average age of a stock could change without any change in the average length, both as a result of increased growth rates and because for adult baleen whales length is not closely correlated with age. Where lengths as such can be most useful is in monitoring populations of a species where growth continues to high ages, such as the Antarctic stocks of sperm whales, almost exclusively mature males that continue to grow after puberty. The length statistics showed a steady decline in the mean length of the catches. Between 1946/47 and 1958/59, for example, the percentage of large sperm whales (> 50 ft) in the catches, declined from approximately 60 per cent to under 20 per cent (Jonsgard, 1960).

The mean length at sexual maturity was early used as an indicator of maturity and despite changed growth rates can be used to calculate the percentage of immature whales in the catches for which only species, sex, and length were recorded (Laws, 1960b). Interpretation of the results may be difficult, for it has been argued that in the absence of other evidence, such as CPUE, an increasing percentage of immatures could be indicative of either increasing recruitment or decreasing numbers of mature whales.

Size has now been largely superseded by other records as indicators of the state of the stocks (age estimates from ovaries, baleen ridges, teeth, and ear plug layering) but has a potential application in the monitoring of unexploited stocks where post-mortem material is not available. Two examples can be mentioned.

Møhl, Larsen, and Amundin (1976) have described an acoustic

approach to determining sizes of sperm whales from the sonar clicks emitted by the species. Norris and Harvey (1972) proposed that the inter-pulse interval during the trains of pulses that make up the clicks is a function of the length of the spermaceti organ and the velocity of sound within it. There are said to be sound-reflecting mirrors—air sacs —at anterior and posterior ends of the spermaceti organ; an estimate of the velocity of sound in it is available and so an estimate of the length of the organ can in principle be determined from the inter-pulse interval and in turn related to body length. In two cases reported by Møhl et al. (1976), which were at the upper and lower ranges of body size, this method successfully predicted body length. It seems that further research could be rewarding.

Whitehead and Payne (1976) have suggested a method by which right whale *Eubalaena glacialis* sizes and ages can be obtained for living animals. They are measured by photogrammetry, using a white circle placed next to the whale from a boat. When viewed from an angle a circle presents a true diameter length and can be photographed from a clifftop or an aircraft. Measured whale lengths and morphological ratios, especially head length : total length were used to estimate age. Known individuals were remeasured for growth rates and calves were measured against their mother. In the absence of absolute length measurements it is claimed that length and therefore age can be estimated from the ratio of snout to blowhole : total length; or snout to dorsal fin : total length. Length frequencies and possibly age structure of other species could be estimated from aerial photographs.

Foetal length measurements may provide another means of monitoring stock condition. Thus Laws (1961) analysed a large number of fin whale foetal lengths representing the period 1925 to 1958. The results were suggestive of an increased growth rate, because the mean size of large foetuses measured in March shows the most definite trend, but they could reflect a long-term change in the mating season.

Changes in the age composition of exploited stocks can now be followed more readily because large and representative samples of the catches are obtained. In the case of whales ovarian corpora albicantia (Mackintosh and Wheeler, 1929; Laws, 1961) and baleen plate ridges (Ruud, 1945; Hylen et al., 1955) were formerly used, but these were restricted respectively to mature females and to younger age groups. Ovarian corpora counts have also been used to study seal age distributions (Bertram, 1940). As indicators of age they have now been replaced by serial layers in tooth dentine or cement (Laws, 1952) or lamina in the ear plugs (Purves, 1955; Purves and Mountford, 1959). Recoveries of long-term whale marks and seal tags, or resightings of branded seals have helped to resolve the question of the rate of deposition of these layers. Although some problems of preparation and interpretation

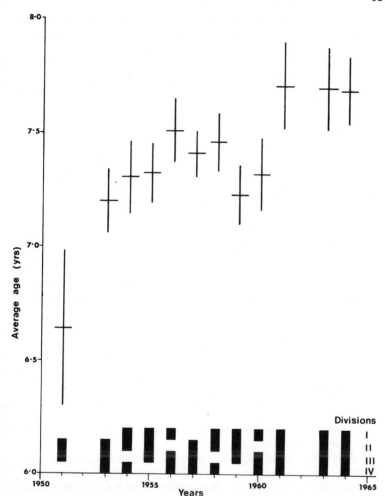

Figure 3 Monitoring the catch of elephant seals *Mirounga leonina* at South Georgia following the introduction of new management measures in 1952. Mean age of annual catches (adult males) and 95 per cent confidence limits. Up to 1960 four sealing Divisions were worked on an annual rotation as indicated; from 1961 all Divisions were involved.

remain they are now used routinely to construct catch curves for whales (Chapman, 1974) and seals (Sergeant, 1976; Lett and Benjaminsen, 1977) and to estimate mortality rates for application to population models and in stock assessments.

In 1952 Laws introduced a management procedure for the southern elephant seal at South Georgia which included routine monitoring of the age composition of the commercial catch which was confined to large males. Because the sealing method progressively and inevitably selected the largest males on the breeding beaches throughout the season, the mean age of the catch provided an estimate of changes in the mean age of breeding bulls in the stock. Figure 3 shows how this age increased, following the introduction of the new management system, to stabilize at about 7.7 years. This implies that a sustained yield situation had been reached and maintained as predicted on the basis of population studies (Laws, 1960a). Had the mean age of the catch begun to increase or to fall the catches could have been adjusted up or down. For economic reasons this industry was tied to the local whaling industry and came to an end in 1964 when whaling at South Georgia became uneconomic due to overhunting.

Although it is so important in studying the dynamics of populations there is surprisingly little information on density dependence in marine mammals. For the northern fur seal Lander and Kajimura (1976) summarize the information on density-dependent mortality of pups, collected over a long period. At low density populations the percentage annual increase in pups born ranged from 6 to 8.5 per cent (table 2). The natural mortality of pups on land was density dependent, but varied widely from year to year and between rookeries; it can be briefly summarized as follows:

Period	Pups born	Percentage pup mortality
1914–1924	143 000	1.7–3.1
1949–1951	451 000	12–17
mid-1950s	461 000	15–20
recent	270 000	c. 10

When density is plotted against pups born the relationship appears to be non-linear.

For grey seals, Coulson and Hickling (1964) demonstrated a relationship between pup mortality (10.5 to 21.1 per cent) and number of pups born per 100 m of accessible shore, that was linear. Continued monitoring of pup mortality (Bonner and Hickling, 1971; Hickling, Hawkey, and Harwood, 1977) has provided further data up to 1976 when pup mortality reached 27.3 per cent for the total population (figure 1). Mortalities above 22 per cent were associated with years after drastic management disturbance involving the killing of breeding cows and— to this extent at least—are unnatural. However, on some individual islands pup mortality exceeded 35 per cent even in the absence of disturbance. The growth rate was also found to be density dependent and presumably influenced survival.

Age and size data can be used to construct growth curves which can themselves provide useful indices of condition. Thus, the exploited elephant seal population at South Georgia showed accelerated growth and earlier maturity (2 years in females) when compared with the stock at Macquarie Island (4 years) (Carrick, Csordas, and Ingham, 1962; Bryden, 1968). Current studies at South Georgia show that the population has adjusted in the intervening 13 years so as to closely resemble the Macquarie Island situation, puberty having been deferred (McCann, British Antarctic Survey, unpublished reports). Another example of this effect is the harp seal, in which sexual maturity is at four years for both sexes in a heavily exploited population, but is deferred by at least a year at higher densities; the effect is more pronounced in the male (Sergeant, 1973).

The mean age at maturity can be established by direct study of the reproductive status of animals in collected samples. If sampling is repeated as in previous examples any changes can be monitored. In the Antarctic fin whale for example, Ohsumi (1972) gave mean ages at maturity from samples collected in 1957, 1961, 1964, and 1968 of 11.5, 10.6, 9.2, and 6.0 years respectively. Lockyer (1972, 1974) extended this method backwards in time when she showed for fin and sei whales that layers deposited in a whale ear plug during immature and mature years show different patterns and the transition between them occurs when the animal becomes sexually mature. A count of the irregular immature layers gives an estimate of the age at maturity. Thus, for current samples of ear plugs the age at sexual maturity can be obtained for each past year-class. This has shown that the year-classes of fin whales born up to 1930 attained maturity at about ten years; from the 1930s onwards this advanced to about six years in the 1950 year class. Sei whales show similar changes. The start of this trend in sei whales preceded the start of large-scale exploitation of this species and supports the idea that it is brought about by increased food availability promoting increased growth.

Laws (1977a, 1977b) reported that a similar transition zone could be detected in the cementum layers of crabeater seal *Lobodon carcinophagus* teeth. This analysis indicated that off the west coast of the Antarctic Peninsula the mean age at sexual maturity was about four years in seals born in the 1930s and 1940s, but decreased from about 1955 onwards and by 1970 averaged about 2.5 years. This seems to be correlated with a rapid reduction of the whale stocks in the same area from 1955 onwards.

A further important indicator of the state of a stock is the adult reproductive rate, usually measured by the percentage of pregnant adult females. It is difficult to establish for seals unless representative pelagic catches are made, and even then there are difficulties because the proportions of pregnant and non-pregnant females may vary geographically. For baleen whales the data are better, though corrections have to be made for the underrepresentation of lactating females in Antarctic waters. Mackintosh (1942) showed that the percentage of adult female blue and fin whales that were pregnant was higher in the pelagic catches from 1932–1941 than at South Georgia in 1925–1931 and suggested that the rate of breeding was becoming faster. Laws (1961) confirmed this increase, using more extensive material from both pre- and post-war years, and suggested that it was related to greater availability of food as a result of the reduction of the whale stocks by whaling. Gambell (1973) reviewed the evidence, making use of yet more recent data, and showed that for fin and blue whales pregnancy rates increased from 25 to 50–60 per cent in both pre-war and post-war periods. Large-scale exploitation of sei whales did not begin until the 1960s, but the increase in their pregnancy rate preceded their exploitation, thus strengthening the conclusion that food availability was the cause.

As well as providing indices of the condition of the stocks these parameters—mortality rate, birth rate, growth rate, and age at first maturity—are all important in the modelling of whale and seal populations discussed earlier.

The future

There is a need to improve knowledge of the biology of individuals and populations, and a continuing need for basic research. After all, two of the most significant discoveries in methods of ageing arose from pure research: the use of internal layers in teeth during the study of a colonizing population of elephant seals (Laws, 1952); the whale ear plug, during a fundamental study of hearing in cetacea (Purves, 1955).

A major problem in the future will be to monitor the stocks of cetacean species that are not exploited directly—a problem that has always existed with most dolphins and small whale species. With the classification by the International Whaling Commission of a number of stocks of large whales as 'protection stocks', the problem becomes pressing, particularly if a large scale krill fishery develops in the Southern Ocean. Two key questions are being posed. first 'are depleted stocks recovering. and if so how rapidly?', and secondly, 'how will this presumed recovery be affected by the development of an industry competing with whales for krill?' These are very practical questions: on the answers depends the future policy for conserving and managing the living resources of the Southern Ocean, because it is not a sufficient safeguard to declare a moratorium on whaling. One important reason is the problem of monitoring the stocks; they could decline further even under a moratorium and a properly managed industry is still the best way of monitoring stocks. Even in the current absence of a major krill fishery, some

WHALE AND SEAL POPULATIONS 135

species may now be adjusting downwards to increased competition from other krill consumers, including other large whale species, seals, birds, fish, and squid (Laws, 1977a, 1977b). If this were to be demonstrated the best conservative management strategy to aim for may well be to manage the ecosystem by harvesting other groups, even if this were uneconomic at the level of individual species. The Southern Ocean has already been seriously perturbed and drastic action, based on firm scientific data may be called for. For example, it has been estimated that Antarctic seals now take 64 million tonnes of krill annually compared with 43 million tonnes consumed by whales (Laws, 1977a); the chicks alone of two species of penguins at South Georgia require one million tonnes of krill annually to reach fledgling size (Everson, 1977).

Also, in explaining changes in abundance it is important not to forget that natural climatic changes can be responsible. This is most likely the explanation for the marked decline in Falkland sea lions and possibly northern sea lions mentioned earlier. Vibe (1967) examined this question in relation to marine mammals in the Greenland area. He recognized three main drift ice stages over the periods 1810–1860, 1860–1910, and 1910–1960 which he claimed were reflected in the catch statistics. He figures changes in the catches of ringed seals, harp seals, and other species which show varying trends of abundance suggesting that they are sensitive to environmental changes. Most interestingly he suggests that changes in the environment may have been responsible for the dramatic reduction in the Greenland whale *Balaena mysticetus* from about 1740 to 1900, 'possibly to a greater extent than whaling'. He gives data on the number of whales caught from 1670 onwards. The preferred summer feeding grounds were above the continental slope where depths were 200–1000 m; when they were covered by pack-ice the whales' access was restricted and they starved. He makes a plausible case for the influence of pack-ice distribution on their decline, especially in the period 1860–1920 when a combination of whaling and this natural factor may have brought them near to extinction. Another long-documented fishery was the Faeroese fishery for pilot whales *Globicephala melaena*; statistics are available back to 1584 (Mitchell, 1973).

Because of the vast areas involved (and therefore the expense), direct observation or counts of living whales may not be the answer. If the counts are carried out by commercial whalers there is always the possibility of bias due to geographical limitations and whalers' species preferences—because the fleets are small and cannot adequately sample the area (36 million km² south of the Antarctic Convergence) even when whales are concentrated for feeding. The cost and effectiveness of non-commercial observers probably rules out purely scientific surveys. On the basis of present knowledge this appears to leave open only a few possibilities—scientific whaling operations, using special ships to sample the populations at intervals, taking a few hundred of each species from each of the unit stocks; or perhaps establishing a series of acoustic buoys to record whale sounds as indices of relative abundance. The former has little advantage over commercial whaling provided the commercial harvests are properly determined and properly monitored. The latter would of course involve overcoming technical (specific identifications) and practical problems (deployment, pack-ice, icebergs), but a feasibility study should be made. It is surprising that so little use has been made of aircraft for counting whale populations; aerial photography, particularly of sperm whales and right whales, but possibly also of rorquals, could perhaps be developed as a means of estimating the ages of living whales as Whitehead and Payne (1976) have attempted. Another possibility is length estimation of sperm whales from sonar clicks (Møhl et al., 1976).

If samples from the catches continue to be available, the maximum information must be obtained from them. The use of back-calculation from earplug layer thickness to establish age at first maturity (and possibly growth rates) has been mentioned. Instantaneous growth rate indices could be established for whales and seals as for elephants (McCullagh, 1969) to give early warning of growth rate changes.

Ideas change and monitoring should not be too limited, because at some future time our present monitoring operations, if they include sufficient measurements, will be invaluable to test hypotheses. There is a need to establish more long-term programmes; most studies tend to be on a short time scale and when terminated after a few years may, despite the investment of time and funds, have contributed little to the monitoring of the population. This emphasizes the importance of standardizing methods so that series of counts carried out by different investigators are truly comparable; when methods are changed it is important to allow for a period of overlap for calibration.

The immediate need is for further coordination of efforts, the continuation and strengthening of projects already in operation, and the establishment of pilot projects to develop new techniques. Methods need to be standardized and calibrated and arrangements for storage and retrieval of data improved because records must be longer than a single worker's career (National Science Foundation, 1977).

References

Allen, K. R. (1977) Updated estimates of fin whale stocks. *International Whaling Commission, Report*, **27**, 221.

Allen, K. R. and Kirkwood, G. P. (1977) Further development of sperm whale population models. *International Whaling Commission, Report*, **27**, 106–112.

Anon (1977a) Annual counts of gray whales at record highs. *Northwest and Alaska Fisheries Center. Monthly Report.* March 1977, 26.

Anon (1977b) Sea lion census conducted off Alaska indicates population decline. *Northwest and Alaska Fisheries Center. Monthly Report,* December 1977, 16.

Anon (1977c) Report of the Advisory Committee on Marine Resources Research. Working Party on Marine Mammals. *Food and Agriculture Organization of the United Nations. FAO Fisheries Report* no. **194**, FIR/R194 (En), 1–43.

Anon (in press) Report of the Advisory Committee on Marine Resources Research, Working Party on Marine Mammals. Appendix B, Proceedings of the Scientific Consultation. *Food and Agriculture Organization of the United Nations.* FAO Fisheries Report.

Bartholomew, G. A. and Hubbs, C. L. (1960) Population growth and seasonal movements of the northern elephant seal, *Mirounga angustirostris. Mammalia,* **24** (3), 313–324.

Bertram, G. C. L. (1940) The biology of Weddell and crabeater seals, with a study of the comparative behaviour of the Pinnipedia. *British Museum (Natural History), Scientific Reports British Graham Land Expedition, 1934–1937,* **1**, 1–139.

Bertram, G. C. L. (1950) Pribilof fur seals. *Arctic,* **3**, 74–85.

Best, P. B. (1977) Status of whale stocks off South Africa, 1975. *International Whaling Commission, Report,* **27**, 116–121.

Best, P. B. and Gambell, R. (1968) The abundance of sei whales off South Africa. *Norsk Hvalfangsttidende,* **57** (6), 168–174.

Best, P. B. and Lurmon, L. C. (1974) Conservation and utilisation of whales off the Natal coast. *Journal South African Wildlife Management Association,* **4** (3), 149–156.

Bonner. W. N. and Hickling. G. (1971) *Grey seals at the Farne Islands, a management plan,* 21 pp. London: Natural Environment Research Council.

Brownell, R. L. (1973) South American fur seal (*Arctocephalus australis*). In *Federal Register, Washington D.C.,* **38** (147), 20572.

Brownell, R. L. (1977) Current status of the gray whale. *International Whaling Commission Report,* **27**, 209–211.

Bryden, M. M. (1968) Control of growth in two populations of elephant seals. *Nature,* **217**, 1106–1108.

Capstick, C. K. and Ronald, K. (1976) Modelling seal populations for herd management. *Food and Agriculture Organization of the United Nations, Scientific Consultation on Marine Mammals, Bergen, Norway.* ACMRR/MM/SC/77, 1–19.

Carrick, R., Csordas, S. E., and Ingham, S. E. (1962) Studies on the southern elephant seal *Mirounga leonina* (L.), 4. Breeding and development. *Commonwealth Scientific and Industrial Research Organisation, Wildlife Research,* **7**, 161–197.

Chapman, D. G. (1974) Estimation of population parameters of Antarctic baleen whales. In *The Whale Problem,* ed. Schevill, W. E. Ch. 14, pp. 336–351. Cambridge, Massachusetts: Harvard University Press.

Condy, P. R. (1977) The ecology of the southern elephant seal *Mirounga leonina* (Linnaeus 1758) at Marion Island. DSc Thesis, University of Pretoria, 1–146.

Condy, P. R. (in press) The distribution, abundance and annual cycle of fur seals (*Arctocephalus* spp.) on the Prince Edward Islands. *South African Journal of Antarctic Research.*

Coulson, J. C. and Hickling, G. (1964) The breeding biology of the grey seal, *Halichoerus grypus* (Fab.), on the Farne Islands, Northumberland. *Journal of Animal Ecology,* **33**, 485–512.

Croze, H. (1972) A modified photogrammetric technique for assessing age structures of elephant populations and its use in Kidepo National Park. *East African Wildlife Journal,* **10** (2), 91–116.

Dawbin, W. H. (1956) The migrations of humpback whales which pass the New Zealand coast. *Transactions of the Royal Society of New Zealand,* **84** (1), 147–196.

Doi, T. (1974) Further development of whale sighting theory. In *The Whale Problem,* ed. Schevill, W. E. Ch. 16, pp. 359–368. Cambridge, Massachusetts: Harvard University Press.

Eberhardt, L. L. (1978) Transect methods for population studies. *Journal of Wildlife Management,* **42** (1), 1–31.

Everson, I. (1977) The living resources of the Southern Ocean. *Food and Agriculture Organization of the United Nations*, GLO/50/77/1, 1–156.

Gambell, R. (1972) Sperm whales off Durban. *Discovery Reports*, **35**, 199–358.

Gambell, R. (1973) Some effects of exploitation on reproduction in whales. *Journal of Reproduction and Fertility, Supplement*, **19**, 531–551.

Gambell, R. (1976) Population biology and the management of whales. In *Applied Biology*, ed. Coaker, T. H. Vol. 1, pp. 247–343. London: Academic Press.

Gulland, J. (1976) A note on the abundance of Antarctic blue whales. *Food and Agriculture Organization of the United Nations. Scientific Consultation on Marine Mammals, Bergen, Norway*. ACMRR/MM/SC/76, 1–11.

Hamilton, J. E. (1939) A second report on the southern sea lion, *Otaria byronia* (de Blainville). *Discovery Reports*, **19**, 121–164.

Heyland, J. D. (1974) Aspects of the biology of beluga (*Delphinapterus leucas* Pallas) interpreted from vertical aerial photographs. *Proceedings of the second Canadian Symposium on Remote Sensing, Guelph, Ontario*, **2**, 373–390. Ottawa: Canadian Remote Sensing Society.

Hickling, G., Hawkey, P., and Harwood, L. H. (1977) The grey seals of the Farne Islands: the 1976 breeding season. *Transactions of the Natural History Society of Northumbria*, **42** (6), 119–126.

Hylen, A., Jonsgard, A., Pike, G. C., and Ruud, J. T. (1955) A preliminary report on the age composition of Antarctic fin whale catches 1945/46 to 1952/53. *Norsk Hvalfangsttidende*, **44** (10), 1–8.

International Whaling Commission (1976) *Report*, **26**, 1–447.

Johnson, A. M. (1976) Population increases and changes in distribution of three species of marine mammals in the North Pacific Ocean. *Food and Agriculture Organization of the United Nations, Scientific Consultation on Marine Mammals, Bergen, Norway*. ACMRR/MM/SC/108, 1–7.

Jonsgard, A. (1960) On the stocks of sperm whales (*Physeter catodon*) in the Antarctic. *Norsk Hvalfangsttidende*, **46** (7), 289–299.

Kenyon, K. W. and Rice, D. W. (1961) Abundance and distribution of the Steller sea lion. *Journal of Mammalogy*, **42**, 223–234.

Lander, R. H. and Kajimura, H. (1976) Status of northern fur seals. *Food and Agriculture Organization of the United Nations, Scientific Consultation on Marine Mammals, Bergen, Norway*. ACMRR/MM/SC/34, 1–50.

Lavigne, D. M., Oritsland, N. A., and Falconer, A. (1977) Remote sensing and ecosystem management. *Norsk Polarinstitutt, Skrifter*, nr. **166**, 1–51.

Laws, R. M. (1952) A new method of age determination for mammals. *Nature*, **169**, 972.

Laws, R. M. (1953) The seals of the Falkland Islands and Dependencies. *Oryx*, **2** (2), 87–97.

Laws, R. M. (1960a) The southern elephant seal (*Mirounga leonina* Linn.) at South Georgia. *Norsk Hvalfangsttidende*, **49** (10), 466–476, (11), 520–542.

Laws, R. M. (1960b) Problems of whale conservation. *Transactions of the North American Wildlife Conference*, **25**, 304–319.

Laws, R. M. (1961) Reproduction, growth and age of southern fin whales. *Discovery Reports*, **31**, 327–486.

Laws, R. M. (1962) Some effects of whaling on the southern stocks of baleen whales. In *The exploitation of natural animal populations*, ed. Le Cren, E. D. and Holdgate, M. W. pp. 137–158. Oxford: Blackwell.

Laws, R. M. (1969) The Tsavo Research Project. *Journal of Reproduction and Fertility. Supplement*, **6**, 495–531.

Laws, R. M. (1974) Population increase of fur seals at South Georgia. *Polar Record*, **16** (105), 856–858.

Laws, R. M. (1977a) Seals and whales in the Southern Ocean. *Philosophical Transactions of the Royal Society, B*, **279**, 81–96.

Laws, R. M. (1977b) The significance of vertebrates in the Antarctic marine ecosystem. In *Adaptations within Antarctic Ecosystems*, ed. Llano, G. A. pp. 411–438. Houston: Gulf Publishing Company.

Lett, P. F. and Benjaminsen, T. (1977) A stochastic model for the management of the Northwestern Atlantic harp seal (*Pagophilus groenlandicus*) population. *Journal of the Fisheries Research Board of Canada*, **34** (8), 1155–1187.

Lockyer, C. (1972) The age at sexual maturity of the southern fin whale (*Balaenoptera physalus*) using annual layer counts in the ear plug. *Journal du Conseil International pour l'Exploration de la Mer*, **34** (2), 276–294.

Lockyer, C. (1974) Investigation of the ear plug of the southern sei whale, *Balaenoptera borealis*, as a valid means of determining age. *Journal du Conseil International pour l'Exploration de la Mer*, **36** (1), 71–81.

Mackintosh, N. A. (1942) The southern stocks of whalebone whales. *Discovery Reports*, **22**, 197–300.

Mackintosh, N. A. (1965) *The stocks of whales*. London: Fishing News (Books).

Mackintosh, N. A. and Brown, S. G. (1956) Preliminary estimates of the southern populations of the larger baleen whales. *Norsk Hvalfangsttidende*, **45** (9), 469–480.

Mackintosh, N. A. and Wheeler, J. F. G. (1929) Southern blue and fin whales. *Discovery Reports*, **1**, 257–540.

McCullagh, K. G. (1969) The growth and nutrition of the African elephant. I. Seasonal variations in the rate of growth and the urinary excretion of hydroxyproline. *East African Wildlife Journal*, **7**, 85–90.

Mitchell, E. D. (1973) The status of the world's whales. *Nature, Canada*, **2** (4), 9–27.

Mitchell, E. D., ed. (1975) Review of biology and fisheries for smaller cetaceans. *Journal of the Fisheries Research Board of Canada, Special Issue*, **32** (7), 889–1242.

Mitchell, E. D. (1977) Evidence that the northern bottlenose whale is depleted. *International Whaling Commission, Report*, **27**, 195–203.

Møhl, B., Larsen, E., and Amundin, M. (1976) Sperm whale size determination: outlines of an acoustic approach. *Food and Agriculture Organization of the United Nations, Scientific Consultation on Marine Mammals, Bergen, Norway*. ACMRR/MM/SC/84, 1–4.

National Science Foundation (1977) Long-term ecological measurements. Report of a Conference, Woods Hole, 1–26.

Norris, K. S. and Harvey, G. W. (1972) A theory for the function of the spermaceti organ of the sperm whale. *National Aeronautics and Space Agency, U.S.A.*, SP-262, 1–9.

Ohsumi, S. (1972) Examination of the recruitment rate of the Antarctic fin whale stock by use of mathematical models. *International Whaling Commission, Report*, **22**, 69–90.

Payne, M. R. (1977) Growth of a fur seal population. *Philosophical Transactions of the Royal Society, B*, **279**, 67–79.

Perrin, W. F., Smith, T. D., and Sakagawa, G. T. (1976) Status of populations of spotted dolphin, *Stenella attenuata*, and spinner dolphin, *Stenella longirostris*, in the Eastern Tropical Pacific. *Food and Agriculture Organization of the United Nations, Scientific Consultation on Marine Mammals, Bergen, Norway*. ACMRR/MM/SC/27, Add. 1, 1–23.

Purves, P. E. (1955) The wax plug in the external auditory meatus of the Mysticeti. *Discovery Reports*, **27**, 293–302.

Purves, P. E. and Mountford, M. D. (1959) Ear plug laminations in relation to the age composition of a population of fin whales (*Balaenoptera physalus*). *Bulletin of the British Museum (Natural History), Zoology*, **5** (6), 123–161.

Reeves, R. R (1977) Exploitation of harp and hooded seals in the western North Atlantic. *U.S. Marine Mammal Commission Report*, no. MMC–76/05, 1–57.

Rice, D. W. (1961) Census of the California gray whale, 1959/60. *Norsk Hvalfangsttidende*, **50** (6), 219–225.

Rice, D. W. and Wolman, A. A. (1971) The life history and ecology of the gray whale (*Eschrichtius robustus*) *The American Society of Mammalogists*, Special Publication, no. **3**, 1–142.

Ruud, J. T. (1945) Further studies on the structure of the baleen plates and their application to age determination. *Hvalradets Skrifter*, **29**, 1–69.

Seber, G. A. F. (1973) *The Estimation of Animal Abundance and Related Parameters.* London: Griffin, 1–156.

Sergeant, D. E. (1973) Environment and reproduction in seals. *Journal of Reproduction and Fertility*, Supplement, **19**, 555–561.

Sergeant, D. E. (1976) History and present status of populations of harp and hooded seals. *Biological Conservation*, **10**, 95–118.

Sergeant, D. E. (1977) Stocks of fin whales *Balaenoptera physalus* L. in the North Atlantic Ocean. *International Commission on Whaling, Report*, **27**, 460–473.

Siniff, D. B., De Master, D. P., Hofman, R. J., and Eberhardt, L. L. (1977). An analysis of the dynamics of a Weddell seal population. *Ecological Monographs*, **47** (3), 319–355.

Smith, E. A. and Burton, R. W. (1970) Weddell seals of Signy Island. In *Antarctic Ecology*, ed. Holdgate, M. W. pp. 415–428. London: Academic Press.

Smith, T. G. and Stirling, I. (1978) Variation in the density of ringed seal (Phoca Lispida) birth lairs in the Amundsen Gulf, Northwest Territories. *Journal of the Fisheries Research Board of Canada*, (in press).

Stirling, I., Archibald, W. R., and De Master, D. (1977) Distribution and abundance of seals in the eastern Beaufort Sea. *Journal of the Fisheries Research Board of Canada*, **34** (7), 976–988.

Strange, I. (1972) Wildlife in the Falklands. *Oryx*, **11** (4), 241–257.

Summers, C. F. (in press) Trends in the size of British grey seal populations. *Journal of Animal Ecology*.

Vaughan, R. W. (1971) Aerial photography in seals research. In *The application of aerial photography to the work of the Nature Conservancy*, ed. Goodier, R. pp. 88–98. Edinburgh: The Nature Conservancy.

Vibe, C. (1967) Arctic animals in relation to climatic fluctuations. *Meddelelser om Grønland*, **170** (5), 9–199.

Watson, R. M., Graham, A. D. and Parker, I. S. C. (1971) A comparison of four East African crocodile *Crocodylus niloticus* Laurenti populations. *East African Wildlife Journal*, **9**, 25–34.

Whitehead, H. and Payne, R. (1976) New techniques for assessing populations of right whales without killing them. *Food and Agriculture Organization of the United Nations, Scientific Consultation on marine mammals, Bergen, Norway.* ACMRR/MM/SC/79, 1–23.

Biological monitoring around an oil refinery

D. H. DALBY
Botany Department, Imperial College, London, England

E. B. COWELL and W. J. SYRATT
The British Petroleum Company, Moor Lane, London, England

The Mongstad environmental studies

This paper is concerned with the approaches used in the biological monitoring studies carried out by ecologists of the British Petroleum Company at the Rafinor oil refinery at Mongstad on Fensfjord, 50 km NNW of Bergen, Norway. British Petroleum, as constructors of the refinery, started a monitoring programme in 1972, and are still continuing the work following the commissioning of the installation in 1975. The background to the study and much other relevant information is to be found in Syratt and Cowell (1975); here it is only necessary to give an outline of the nature of the data collected and how these data may be analysed.

The base-line survey

Many environmental studies require the preparation of a base-line data set as a standard for assessing subsequent changes. The longer the period for which this base-line data set is available the better. It is unusual for such industrially related studies to exceed about three years, but this is a bare minimum to document the scale of the natural fluctuations that may be occurring. Some biologists query the concept of stability in the long term, because of climatic changes, slow responses in the temperature of the ocean waters, etc., quite apart from all the permutations and combinations of species following from interspecific competition and local succession on a rocky seashore.

The Mongstad base-line survey, which incorporates data collected over five years (together with a few sites remote from the refinery which were incorporated later), is not uniform in its treatment of sites. Two (Mongstad sites 12 and 13) were inaccessible and could not be recorded every year, and the sites Mongstad 5 to 10 were destroyed by rock dumping during construction. Consequently, the base-line data set varies in the number of years each site has been sampled, although the

142   D. H. DALBY, E. B. COWELL, AND W. J. SYRATT

sampling itself has followed a standardized procedure (described by Syratt and Cowell, 1975).

Data collection

Some 26 recording transects were established in or near the refinery (see figure 1), and each was studied during a period of low spring tides early in May, ensuring some degree of temporal comparability from

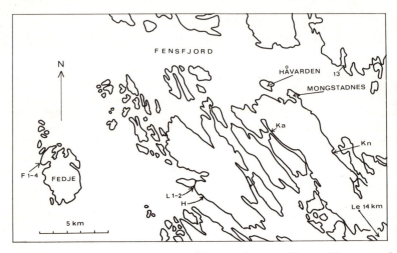

Figure 1 Map of monitoring sites around Mongstad, western Norway. The names of the monitoring sites are abbreviated as follows: F, Fedje; L, Lerøy; H, Hoplandssjøen; Ka, Kalandssjøen; Kn, Knarvik; Le, Leivestad. All the Mongstad sites (1–16) are located on Mongstadnes, apart from 13 which is on the north side of Fensfjorden. The site named Leivested is actually near the head of Rylandspollen, 2.5 km north of the village of Seim in Lindås herred.

year to year. At each site a transect line was taken at right angles to the shore, and species-abundance scores recorded at decimal divisions of the 1.8 m tidal range and extending upwards to the supralittoral fringe and the splash zone. Forty-five species were recorded, their names being given in Syratt and Cowell (1975) and Dalby *et al.* (in press). The species selection is that used in littoral studies in Milford Haven, South Wales, with some modifications due to the more northerly latitude of the Norwegian site. Some new species have been incorporated into the Mongstad records as these have been continued from year to year.

At each of the 10 (or up to 15 in some cases) intertidal stations per transect, the abundance of each species on the list was estimated using the scale of Crisp and Southward (1958), modified by Crapp (1971). This employs a ranking scale in which the abundance measure increases very approximately logarithmically with rank, the maximum score of

7 being the maximum abundance possible in nature for a given species. A rank of say 3 does not imply identity in numbers or biomass between species, but it does imply an approximate similarity in the proportion of the total possible which has been actually found for each.

In addition to the species scoring, additional information has been obtained regarding the general health of the site, relevant changes on shore nearby, and in some instances data on, for example, age structure in selected species, notably the limpet *Patella vulgata*. As discussed later, this peripheral information may prove invaluable in explaining the causes of changes detected in the routine studies.

Problems in detecting change

Only rarely are biological records uniform from year to year and change is the rule rather than the exception. The main problem is not so much in detecting change as in assessing its significance and in attributing a cause. The statistical significance of observed change requires the data to be in a suitable form, and, lacking this, one has to make subjective decisions based on experience and upon observations from comparable reference (control) sites, whilst the attributing of cause may take us far beyond the recorded data into areas of physiological experiment and fundamental research. The following discussion is centred on our interpretation of the Mongstad data rather than on theoretical generalizations.

The Mongstad abundance records employ ranks for each species, so they cannot be used to obtain direct absolute measure of change. However, since successive grades are approximately logarithmically related, we have assumed that the scores can in fact be handled according to the normal rules of arithmetic (and the results of the multivariate studies certainly show that this assumption is not seriously misplaced).

It is generally accepted that one observer may award a rank one above or one below that given by another, assessing exactly the same area of shore, but this will only be a problem in marginal cases where the true score does indeed lie close to the boundary between the categories. It becomes more of a problem however when working with large areas, or if there should be any difficulty in locating the exact study area on another occasion. Changes of 2 (and even 3) units are very common in the Mongstad data during the period of the base-line survey, so we assume that changes of this magnitude may often have natural causes, though some may deserve closer attention if they form part of a sustained trend or are part of a consistent pattern.

The three-dimensional data base (species × stations × transects) allows us to extract an estimate of change in vertical range in the littoral for each species at each site by comparing the number of stations at which each species is present. Because of difficulties in locating the sampling stations within each transect with complete precision from year to year, these figures cannot be more than estimates, and in any case we usually lack information on abundances between stations, but taken in conjunction with changes in abundance they certainly point to sites or species where major changes have occurred.

When we determined abundance changes, it became necessary to simplify the data matrix to a two-dimensional form (species × sites), and this was done by extracting the maximum score for each species at each site. This procedure has several advantages since it can (if need be) be undertaken fairly quickly by hand from the data sheets, it gives a measure of the species' optimal performance at each site, and it provides one set of site descriptors for use in site ordinations using multivariate statistical techniques. Change in abundance was thus calculated by subtracting the maximum score for a species in one year from the maximum score for that species in the following year.

These two sets of calculations, namely changes in number of stations where a species is present, and changes in maximum abundance, were carried out using a specially prepared computer program (MONGT, written in FORTRAN IV), which operates on the complete base-line data matrix stored on permanent file. An example of the output from this program is given in table 1, relating to site Mongstad 15, where there is an established effect on littoral organisms picked up during the 1977 survey. The sum of absolute changes in maximum score greater than 3 between 1976 and 1977 is 25; the sum of absolute changes in number of stations greater than 3 for these same years is 21. An arbitrary value of 15 for both of these sums seems reasonable for alerting us to closer examination of the data, this value being selected after considering data from the base-line during the pre-commissioning period. Major changes are then seen in *Enteromorpha* spp. (max. score up 6, stations up 3), *Pelvetia canaliculata* (down 4, down 2), *Corallina officinalis* (up 4, up 2), *Balanus balanoides* (down 2, down 4), *Littorina saxatilis* (up 3, up 5), *Littorina littoralis* (down 7, down 6), and *Patella* spp. (down 4, down 3). Not all these changes can be easily explained of course, but it is immediately clear that the sudden rise in *Enteromorpha* is accompanied by a fall in *Littorina littoralis* and limpets, and barnacles have also become seriously restricted. We cannot say that these changes are statistically significant—only that the total amount of change is definitely greater than at the majority of biologically equivalent sites in the area.

If we seek a cause for these changes, we need to compare them with changes in similar sites elsewhere at Mongstad, and perhaps also to look at any peripheral information available. Comparison with similar reference sites is important, but this requires some measure of similarity

Table 1 Example of computer output summarizing changes in species abundance and range at each station in the Mongstad survey. The entries for species 12 to 36 have been omitted for clarity. Blanks indicate that a species is absent from both years in a comparison, a dot indicates that the species was present in both but did not show any change

	Maximum scores			Number of stations		
Mongstad 15	1974	1975	1976	1974	1975	1976
	1975	1976	1977	1975	1976	1977
1 Grey green lichens	.	.	−2	−1	+2	.
2 Orange red lichens	.	−1	−1	−2	+2	.
3 Verrucaria maura	+1	.	.	−2	+1	+1
4 Verrucaria mucosa	.	−1	−1	−1	−1	.
6 Lichina confinis	..	+1	−1	−1	.	.
7 Codium fragile						
8 Ulva lactuca						
9 Enteromorpha spp.			+6			+3
10 Alaria esculenta						
11 Pelvetia canalic.	−1	+1	−4	−2	+1	−2
37 Patella spp.	+2	+1	−4	+1	−1	−3
38 Nucella lapillus	−1	.	−2	−3	−1	.
39 Mytilus edulis	−2	−1	+1	.	−1	+1
40 Spirorbis spp.	−2	.	−1	+1	.	+1
41 Pomatoceros triq.	−2	+3	−1	−1	+1	.
42 Ascophyllum free						
43 Fucus serr. free						
44 Fucus vesic. lin.						
45 Fucus dis. anceps						
Sums of changes	−6	−1	−16	−14	+6	+2
Sums of absolute changes G.T.3	0	4	25	0	0	15

between sites. For this we need to understand the main environmental factors at work on the shores, and this goes far beyond our basic data set. Field observation suggested that the main factor is actually exposure to wave action, since rock type, tidal range, etc., do not vary very much locally.

Detecting the wave-exposure gradient

The reduced data matrix (species maximum scores × sites) was subjected to principal components analysis using correlation coefficients as a measure of similarity, and by reciprocal averaging (Hill, 1973) to give site and species ordinations.

Principal components analysis. The first and second components' weightings are used to plot the positions of the sites in figure 2.

The first axis is clearly a measure of wave exposure, since the sites

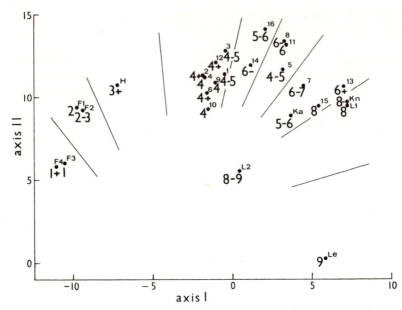

Figure 2 Principal components plot for the Mongstad sites, using loadings on the first two axes. The large numbers are the exposure grade values, whilst the small numbers are the reference numbers for sites on Mongstadnes. Names of other sites, further from the refinery, are abbreviated as in figure 1, which also gives their positions. The inclined lines (inserted by eye only), indicate very approximately the spread for each exposure grade unit on the curved scatter of points, and they have no theoretical justification.

from the island of Fedje on the outer edge of the skerries come, together with the headland at Hoplandssjøen, to the left of the diagram, whilst the sites from the very sheltered bays at Lerøy 1, Knarvik, and Leivestad come to the right, but this gradient appears to be somewhat inclined in relation to axis I. Axis II, as often happens with non-standardized quantitative data, is a measure of species richness with species-poor sites coming at the bottom and species-rich sites at the top. Species richness is of great biological importance, especially since species diversity indices are often used as a measure of environmental stress (see Regier and Cowell, 1972 for this, in the context of pollution studies). In the present study we cannot use absolute diversity values because we are looking at only a small part of the total flora and fauna, and low values for some species might merely mean that others, not recorded, have taken their place. We have, however, listed most of the larger plants and animals, and it does look as though shores of maximum shelter or maximum exposure support a smaller number of species than those of average conditions (in the proportion of about 1 : 2 for our data set). There is also some slight evidence that very exposed shores have rather more species than very sheltered ones, so perhaps some other stress factor is also operating—possibly salinity.

Reciprocal averaging analysis. The multivariate technique of reciprocal averaging has the particular advantage of producing simultaneous ordinations of individuals and their attributes (here sites and species). To avoid confusion these are plotted separately in figures 4 and 5. These two plots show conclusively that the first axis is a wave-exposure gradient, with the most seaward sites and species of known exposure tolerance (e.g. *Fucus distichus* subsp. *anceps* and *Alaria esculenta*) coming at one extreme, and sheltered sites in bays together with shelter-demanding taxa (such as the unattached fucoid algae) coming at the other extreme.

The main conclusions to be drawn from these multivariate analyses are that the principal environmental factors operating at Mongstad, as determined from our base-line survey, are those linked to exposure to wave action, and that if additional sites are examined they should be chosen so as to increase our knowledge of the flora and fauna at the exposure extremes. Two of the sites on Fedje island (3 and 4) and that at Leivestad, were in fact added after the first analyses for this reason, but they are included here to demonstrate how clear the picture has become when almost the whole exposure range is included in the study.

Wave-exposure scale for Mongstad

With the demonstration, by the ordination studies, that the main environmental factor separating the sampling sites at Mongstad is exposure to wave action, we needed to quantify this so that sites of numerically similar exposure could be compared with each other. It must not be thought, however, that wave exposure is a single factor. In reality it is a complex of many components which interact in such a manner as to make it virtually impossible to isolate any single one in the field. Since we were unable to make physical measurements that were sufficiently meaningful, we turned to a biological exposure scale, adapting the scale used by Ballantine (1961) in Pembrokeshire, S. Wales. The details of our scale are given in Dalby *et al.* (in press), together with a discussion of the nature and value of biological exposure scales in general.

In summary, our Mongstad scale involves the preparation of species-performance curves using the maximum abundance attained by each species at each site, plotted against exposure grade. By use of a recursive method of curve fitting and reassessment of exposure grades we calculated adjusted values for each site on a scale from 1 (most exposed) to 9 (most sheltered). These grade numbers on the exposure scale are actually rank positions, with no assumption being possible (or even likely) that they measure even steps along the gradient. There is evidence (presented in Dalby *et al.*, in press) that the grades may be a logarithmic transformation of the actual exposure components, but this is not assumed in the computer program fitting fourth order polynomials to the species abundance data. Whilst we believe that the numerical exposure grades are part of a continuous sequence with decimal values between the unit grades, in practice we think it wholly unjustified to be more precise than to use the nearest quarter unit.

Testing the exposure scale

We have no absolute standards against which to compare our exposure scale, although this is a subject which we are studying further now. We therefore have to use indirect evidence in judging its success. This evidence is of three kinds: agreement with biological relationships established elsewhere, similarity to rankings using multivariate techniques, and coherent patterns using physiographic data from maps.

The biological evidence is discussed more fully by Dalby *et al.* (in press); in brief, it consists of showing that in Pembrokeshire there is a linear relationship between exposure grade and length/aperture ratio of the shell in *Nucella lapillus* (Crothers, 1973), and between the reciprocal of the exposure grade and the height of the top of the black lichen zone dominated by *Verrucaria maura*. The Mongstad sites conform closely to those in Pembrokeshire, with a correlation coefficient for exposure against shell shape ratio of 0.921 ($N = 24, p < 0.001$), and for the reciprocal of exposure grade against lichen height of 0.966 ($N = 23$, $p < 0.001$). We may thus assume that the Mongstad exposure estimates are comparable to those originally defined in Pembrokeshire by Ballantine (1961). In a series of papers (see Crothers, 1973, 1977, etc.) it has been shown that this is true for *Nucella* in many parts of north-west Europe.

The first axis of the principal components analysis was assumed to reflect the wave-exposure gradient; this is now seen to conform with the pattern of exposure grades shown in figure 2, where there is a fairly uniform progression in exposure values in an arc from left to top to right. Only the site at Lerøy 2 (an almost vertical cliff rising from a horizontal silt flat) seems to depart seriously from this arc—almost certainly due to its reduced species complement.

The reciprocal averaging analysis also arranges sites along the first axis according to the principal gradient present in the data, but experience suggests that it does so with greater success. The site spacings on axis I introduce some distortion, but if we accept that this axis can be equated with wave exposure we would expect a high correlation between site exposure grade and rank on axis I. Spearman's rank correlation coefficient proves to be 0.926 ($N = 26$, $p < 0.001$), rising to 0.950 ($N = 25$, $p < 0.001$) if the very aberrant site at Lerøy 2 is omitted.

Finally, in figure 3 we present a map of the Mongstad refinery

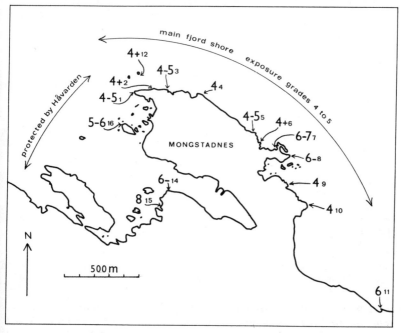

Figure 3 Map of Mongstadnes showing exposure grade values for the monitoring sites 1 to 16 (excepting 13). Shelter is provided by the island of Håvarden for sites 14 to 16, whilst local protection for 7 and 8 was given by the headland and nearby small islets (this area is now much modified, and site 8 is destroyed). Site 11 is on a small reef running out from and almost parallel to the shore.

peninsula to show how the numerical values for the site exposure grades make sense in relation to local topography and fetch. All the lowest scores (between 4 and 5) lie along the main shore of Fensfjord, and those with the highest scores are in bays with local shelter. The maximum fetch lies along Fensfjord to the WNW and ESE of Mongstadnes, but considerable shelter is afforded by the island of Håvarden, together with a number of small islets and shallow water nearby.

Matching Mongstad 12 with other sites

The method of estimating exposure grades thus allows us to quantify the main environmental factor distinguishing sites in our data, and so can be used for locating similar sites for reference purposes.

Returning to the site at Mongstad 15, this has an exposure grade of 8.00, and is the most sheltered on the Mongstad peninsula. The most similar sites in the study area are Lerøy 1 (also 8.00) and Knarvik (7.75). The changes for the seven species referred to earlier are given in table 2 for these three sites. From this table it is clear that the most important

Table 2 Summary of changes in abundance and range for selected species at Mongstad 15 and two other sites of similar exposure grade between 1976 and 1977. The species selected are those with the largest changes at Mongstad 15

	Maximum abundance			Number of stations		
	Mongstad 15	Lerøy 1	Knarvik	Mongstad 15	Lerøy 1	Knarvik
Enteromorpha intestinalis	+6	−2	0	+3	−3	0
Pelvetia canaliculata	−4	0	−1	−2	−1	0
Corallina officinalis	+4	0	−3	+2	0	−1
Balanus balanoides	−2	+1	0	−4	+1	−3
Littorina saxatilis	+3	+2	0	+5	+5	+2
Littorina littoralis	−7	0	−2	−6	0	0
Patella vulgata	−4	0	−1	−3	0	0
Sums of absolute changes greater than 3	25	5	0	15	9	0

changes are for *Enteromorpha* spp., *Littorina littoralis* and *Patella vulgata*, the first having increased and the other two decreased. Apart from *Corallina*, all other changes can be matched by similar-sized changes outside the immediate refinery area and so can be ignored. Knowing the effects of algal grazing by winkles and limpets, we can assume that between May 1976 and May 1977 something happened at Mongstad 15 which adversely affected these gastropods and so favoured the establishment of the green algae.

Whatever the causes were, their effects were extremely local and virtually confined to the small bay in which Mongstad 15 is situated. The adjacent site (Mongstad 14) shows virtually no changes at all. The changes in *Corallina* were more widespread (up 4 at Mongstad 14, up 1 at Mongstad 16), and are probably not related to the localized modifications at Mongstad 15.

At the time of the survey in May 1977 we saw several signs of local contamination, mainly in the form of oil on the water surface in the bay at Mongstad 15, and the presence of sewage fungus in the small stream entering the bay. The biological data then suggest that effluents in the stream were the cause of most of the changes observed, and enquiries showed that there had in fact been a fractured pipe within the refinery processing area that had led to a temporary contamination of the stream (see Monk, Cowell, and Syratt, 1978).

General conclusions

Some conclusions can be drawn from the contents of the present paper which have general relevance to marine pollution. They are discussed briefly below.
1. If biological monitoring is employed, recording should be quantitative and in such a form that the data can be treated statistically. In the present study much was achieved, without the application of statistical tests, by concentrating on the largest changes, but even more information could have been extracted with a different kind of data recording (which would certainly have been more time-consuming and expensive).
2. Variations in species abundance on rocky shores are often considerable from year to year, and these must be allowed for in assessing the magnitude of possible contamination.
3. No monitoring study of this kind should be carried out in a vacuum without a fairly full understanding of the general ecological processes taking place in the natural environment.
4. Biological data become more useful when the species studies are employed in a kind of bioassay; this means that we will benefit from restricting the recording to rather few species of known tolerance and response. Against this we do not necessarily know what these tolerances and responses are, nor what forms the possible pollution may take. One is not however working wholly in the dark here, as there is a large amount of information available on laboratory and field responses by littoral organisms exposed to oil industry pollutants, and the species selection should be influenced by this experience.

Acknowledgement
The authors gratefully acknowledge help from Rafinor A/S, the Norsk Hydro Company, Dr A. J. Morton, Mr R. G. Davies, and the advisory staff of the Imperial College Computing Centre. We also thank Tor Bokn of N.I.V.A., Oslo, for valuable discussions in the field.

References
Ballantine, W. J. (1961) A biologically-defined exposure scale for the comparative description of rocky shores. *Field Studies*, **1**, 1–19.
Crisp, D. J. and Southward, A. J. (1958) The distribution of intertidal organisms along the coasts of the English Channel. *Journal of the Marine Biological Association of the United Kingdom*, **37**, 157–208.
Crothers, J. H. (1973) On variation in *Nucella lapillus* (L.); shell shape in populations from Pembrokeshire, South Wales. *Proceedings of the Malacological Society, London*, **40**. 319–327.

Crothers, J. H. (1977) On variation in *Nucella lapillus* (L.): shell shape in populations from the Channel Islands and north-western France. *Proceedings of the Malacological Society, London*, **43**, 181–188.

Dalby, D. H., Cowell, E. B., Syratt, W. J., and Crothers, J. H. (in press) An exposure scale for marine shores in western Norway. *Journal of the Marine Biological Association of the United Kingdom.*

Hill, M. O. (1973) Reciprocal averaging: an eigenvector method of ordination. *Journal of Ecology*, **61**, 237–249.

Monk, D. C., Cowell, E. B., and Syratt, W. J. (1978) *The littoral ecology of the area around Mongstad refinery, Fensfjorden, during the three years after refinery commissioning, 1975–1977.* London: The British Petroleum Company.

Regier, H. A. and Cowell, E. B. (1972) Applications of ecosystem theory, succession, diversity, stability, stress and conservation. *Biological Conservation*, **4**, 83–88.

Syratt, W. J. and Cowell, E. B. (1975) *The littoral ecology of the area around Mongstad refinery, Fensfjorden.* London: British Petroleum Company.

Monitoring radioactivity in the marine environment

N. T. MITCHELL
*Ministry of Agriculture, Fisheries and Food,
Fisheries Radiobiological Laboratory,
Lowestoft, Suffolk, England*

Introduction

There is by now a considerable amount of literature on the distribution of radioactivity in the marine environment. Although this is a topic of world-wide interest, it has been dominated by the UK, in part a reflection of the importance which the marine environment holds for the UK as a means of securing the safe disposal of radioactive wastes from the generation of nuclear power. This paper is therefore confined to an account of methods used in the UK to monitor the effects of controlled radioactive waste disposal on the marine environment. Monitoring occupies an especially important role within the control structure for ensuring the safe disposal of radioactive waste, and its treatment here will be widened to explain this role and the use to which it is put. Although outside the scope of this symposium, the system of monitoring described here applies equally to environments other than the sea; however, it has a special relevance to the seas around the UK because so many of the nation's major nuclear establishments which need to dispose of low-level radioactive waste are in coastal situations and release wastes into the sea, either directly or via estuaries.

The aims and objectives of environmental monitoring; some basic principles

The word monitoring, like radioactivity itself, has become evocative and its meaning and interpretation as applied to radioactivity varies. In the UK and reflecting the emphasis of national waste disposal policy (Anon, 1959) which is on limitation of radiation exposure of the public rather than environmental resources, monitoring has come to be connected primarily with measurements needed to assess doses to which the human populations are exposed. The justification for this is found in the realization that the potential risk to environmental resources is minor (IAEA, 1976) provided that the exacting standards for control of public radiation exposure are maintained (ICRP, 1977).

153

Despite the emphasis on public radiation exposure and its assessment, the objectives of environmental monitoring programmes as practised in the UK are more complex and frequently serve a number of purposes. In addition to the estimation of public radiation exposure within the radioactive waste disposal control structure, a further objective of monitoring, also related to surveillance of the operation of the nuclear establishments, is to provide a means of checking the adequacy of control measures such as chemical treatment plant or filtration systems. Monitoring is often closely allied to research results and together they feed back into the control system and provide evidence to substantiate or modify measures previously taken. Finally, mention must be made of the need to provide public information, an objective which should not become an end in itself and require production of data isolated from those generated for other purposes, but should in most cases be met whilst fulfilling these other purposes.

Monitoring for estimation of public radiation exposure

Though the primary standards for radiological protection of the public are expressed in terms of exposure to radiation rather than concentrations of radionuclides in specific materials, it is so rarely possible to measure human radiation exposure directly that for all practical purposes opportunities for monitoring of this kind can be ignored. Indeed, it is only possible to measure radiation exposure in units of dose at all when the source is external and even then conversion of the environmental measurement (e.g. dose rate in air) is needed to calculate the dose received by man. For all cases of internal radiation exposure it is necessary to calculate doses from monitoring the concentrations of radionuclides in the ingested materials; the work of the International Commission on Radiological Protection (ICRP, 1959) has long been invaluable in this field.

All environmental monitoring programmes depend to some extent on a process of modelling of the environment, though it is in the assessment of public radiation exposure that it has become most highly developed. The modelling system developed in the UK is the critical path approach, which is also used as the basis for the control system (Preston and Mitchell, 1973) in setting authorized limits; this depends on achieving an overall empirical understanding of the way in which radionuclides behave after release to the environment, in particular the routes or 'pathways' through which contamination of environmental materials and human radiation exposure occurs. Identification of these pathways is the first stage of a study known as a 'habits survey' in which the characteristics of exposed populations are identified; this leads to a decision as to which of the pathway(s), material(s), and the individual(s) or group(s) of the public are 'critical' in relation to potential radiation exposure and on which control limits on discharges are set. Most of control monitoring is concerned with critical pathways, the logic of this decision being realization that if exposure through them is controlled within acceptable limits, so too will be doses via non-critical pathways. The emphasis to date has been on exposure of individuals since this, rather than the collective dose to large groups of the public, has been found to be the limiting constraint in every case. Nevertheless, it is necessary to be able to assess collective (population) doses for some purposes, and analysis along exposure pathways provides the model for estimation of collective dose (Jefferies, Mitchell and Hetherington, 1977) to the whole of the population exposed to the effects of a specific disposal—which in an extreme case would be the worldwide population.

Whilst the environmental model describes a number of stages in the pathway, monitoring for the purpose of estimating public radiation exposure will almost invariably concentrate on the penultimate stage in the chain, with most of the effort being devoted to sampling of the critical material itself or measuring the ambient radiation dose-rate in the case of an external exposure pathway. The programme of measurement can usually be restricted to a small number of points in one, or at most only a few, pathways; further details will be found in the examples quoted in section 4.

Monitoring for other control purposes

All important discharges—those sufficient to generate significant public radiation exposure—are under constant surveillance as a result of monitoring of critical pathways. However, this is not the only role of environmental monitoring in our waste disposal control system; it is aimed not merely at limiting exposure within the specified upper limits of the national waste disposal control policy, but also at reducing it to levels which are as far below these limits as reasonably practicable. Monitoring is therefore also designed to provide a means of detecting avoidable releases; although this is mainly achieved by careful sampling and analysis of the radioactive waste discharges themselves, environmental monitoring procedures can be used to discover releases made unwittingly. In these cases the materials collected will not necessarily be any of those which feature in the critical pathway(s). The particular property sought in materials used for this purpose will be that of concentrating one or more of the component radionuclides of the discharge to a high degree in order to facilitate the detection in the environment of effects of the discharge with the greatest possible sensitivity. A non-critical indicator material will often prove to be the most useful.

Research

Environmental research to provide information which enables the consequences of future discharges to be predicted, and the effect of past discharges to be better understood, is an important component of the overall UK waste disposal control strategy. Even if the term monitoring is confined to those measurements which are organized as a regular routine, it is not easy to draw a firm dividing line, for such monitoring may simultaneously make a useful contribution to research; it is undoubtedly true that many research programmes provide a wealth of useful data to supplement those from basic monitoring programmes. The exact distinction drawn between the two becomes somewhat arbitrary, for in the experience gained at FRL it has often been found that a single marine sampling programme can often serve both purposes. For example, sampling of *Porphyra* seaweed, the basic raw material from which the foodstuff laverbread is manufactured, was a routine monitoring requirement in relation to Windscale discharges for many years, its consumption providing a critical pathway on which control of these discharges was based (Preston and Jefferies, 1967). The degree of public radiation exposure was established from the analyses of seaweed, collected along the local Cumbrian beaches. With the addition of data on discharges the same, originally monitoring, data provided the means of deriving an empirical relationship between discharge rate and mean concentration in the seaweed, and this has provided a means of predicting the effect of future discharges. Used in this way the environmental measurements may be regarded as research, just as they are when used in conjunction with analyses of the seawater from which activity in the seaweed has been derived to calculate concentration factors. The derivation of concentration factors for a wide range of radionuclides and marine materials has been a most important feature of the FRL research programme over the last two decades. *Inter alia*, the bank of data that we now hold greatly facilitates the job of assessing levels of radioactivity in the marine environment and in many cases renders the measurement of radionuclides in particular materials unnecessary, for they can often be assessed with adequate accuracy if the activity distribution in the water is known.

The interpretation of data from monitoring programmes

For monitoring programmes destined to be used to estimate public radiation exposure the method of data interpretation is inherent in their design. Data from the habits survey which identified the critical pathway (and hence the critical material being sampled) are used to convert levels of contamination or ambient dose-rates into rates of exposure.

This method applies equally to individuals, to critical groups, and, by extension of the environmental model, to large populations—the main difference being the value adopted for consumption rate or occupancy, although the geographical distribution of large populations is likely to be so wide that their exposure cannot be based on one single value of contamination level in a material or the dose-rate at a single location.

In this way the environmental 'exposure pathway' model which has been the basis for control, including the design of the monitoring programme, can be utilized to evaluate the significance of any series of measurements, comparing the calculated rate of exposure with the primary standards, the ICRP-recommended dose limits. Implicit in the use of this method is the need to go back to first principles each time, but in practice more frequent use is made of a secondary standard known as the derived working limit (DWL), calculated from ICRP data (Preston, 1971). The DWL for internal exposure is, strictly, a tertiary standard since it is calculated from rates of intake of radioactivity (already a secondary standard) which would result in radiation exposure at a rate equal to the ICRP-recommended dose limit.

The DWL is a very convenient working tool and has been used widely throughout the UK. For the purpose of assessing the significance of monitoring data it may be compared with individual measurements, though a more common and meaningful practice is to use annual averages of contamination level or dose-rate, so reducing inevitable fluctuations to manageable proportions and computing a value—the fraction of the ICRP-recommended dose limit—which is consistent with the minimum period (one year) to which the recommendations themselves refer.

More complicated environmental models are needed to deal with situations after isolated events, such as nuclear explosions or reactor accidents. It is necessary for such models to represent the dynamic behaviour of the environment and, in the most complex cases, the mathematical techniques of systems analysis may be required. If sufficient data about the environment can be obtained or postulated, it will still be possible to relate the transient concentration in a compartment of the model to the dose commitment to members of the critical group. For accidental releases of materials that move rapidly through the environment, the maximum concentration in a suitable compartment, may be the most convenient quantity to relate to exposure, an approach which has been used in the planning of environmental monitoring programmes following accidental releases (Bryant, 1971).

In principle, the use of the DWL can also be applied to measurements of an indicator material, i.e. to a material not in a critical pathway. In practice, however, this requires a high degree of realism in the environmental model. An alternative approach is analogous to that used by the ICRP in relation to individual monitoring for internal contamination (ICRP, 1968). In the environmental situation, this method involves selecting a fraction, perhaps a few per cent, of the dose limit to the critical group and calling this an Investigation Level. The environmental model is then used to estimate the value for the indicator measurement corresponding to this investigation level. Since this value depends not only on the selection of the investigation level but also on the environmental model, it may be termed a Derived Investigation Level. As long as results in the indicator material remain below the derived investigation level, the simple environmental programme can be continued. If, however, results rise above the derived investigation level, the reason must be sought and, if necessary, a more conventional monitoring programme aimed at assessing doses must be brought into effect. Although little deliberate use of this concept has been made in the UK, the emphasis having been on the DWL, intuitive use is made of it, with rather arbitrarily chosen values, to discard trivial results.

Monitoring in practice

Selection of monitoring materials

Within the whole range of activities required in surveillance and control of the UK nuclear power programme there are few marine materials that have not been monitored at one time or another and many feature routinely because of a special importance as critical or indicator material.

Internal exposure pathways. In many cases fish or shellfish feature prominently, the particular species which is critical depending not only on the nature of the available stocks but on the radionuclides involved and their biological characteristics. Such factors as these account for shellfish (oysters) being the critical material for discharges from Bradwell rather than fish, and fish rather than shellfish for discharges from Windscale, even though both fish and shellfish are caught in both areas. For similar reasons the most abundant radionuclide in the discharge will not necessarily be critical and, again using Bradwell as the example, the critical radionuclide zinc-65 makes up less than one per cent of the total radioactivity discharged. Fishermen and their families frequently form the critical groups which almost without exception are the local populations living relatively close to the area of the sea where releases are made. The classic exception to this is the *Porphyra*/laverbread pathway which for so long was the critical pathway for Windscale discharges, the exposed population in this case being some 300 km away in South Wales where the foodstuff is eaten. It still exists as a possible pathway but only of potential importance, for harvesting of the seaweed from the Cumbrian beaches near to Windscale began to decline about 10 years ago and finally ceased in 1972. From time to time supplies of seaweed reach South Wales from Walney Island where concentrations are much lower than sites close to Windscale; the levels of radioactivity in seaweed harvested from beaches further away are lower still. As a result, public radiation exposure through this pathway is much less than that through at least two others, fish/shellfish consumption and the external exposure pathway from time spent on areas of the foreshore.

External exposure pathways. The most common of these pathways is where radioactivity is taken up onto sediment and thereby produces a source of radiation to people who frequent the foreshore where sediment lies exposed by the receding tide. The importance of these areas in terms of public radiation exposure will depend on the extent to which they are used, so that once again it is often fishermen who form the critical group. A further factor is the level of contamination of the sediment and thus of the radiation field above it; it is generally found that uptake is greatest on fine particulate sediments such as silts (as opposed to sand) and the highest dose-rates are found where such sediment collects, generally in estuaries but occasionally in enclosed parts of the open coast such as harbours and rocky clefts.

There are other external exposure pathways but the only one of particular note is where fishermen are exposed to radiation when they handle fishing gear, the mechanism being that radioactivity taken up by particulate matter suspended in the sea becomes trapped in the fibres which make up nets and lines. In contrast to the beach-occupancy pathway where gamma radiation is the predominant source, almost all the significant exposure from handling fishing gear comes from beta radiation.

Indicator materials. The most useful indicators in the marine environment are algae and molluscan shellfish (Pentreath, 1976). Exceptionally, concentration factors as high as 10^5 may be exhibited (defined as 'Concentration in the material/Concentration in the water') as shown by zinc-65 in oysters (Preston, 1966). Values for seaweeds vary with species but several of the commonly encountered metal radionuclides are concentrated by factors in the range of 10^3 to 10^4 and this, together with the ease with which they can be found and collected around British coasts, makes them a firm favourite for indicator purposes.

Analytical demands

A wide range of radionuclides occurs in the various waste discharges that are made from nuclear sites in the UK and FRL is called upon to measure many though not necessarily all of them. The methods used have been discussed elsewhere (Dutton *et al.*, 1974) and discussion here will be confined to a brief summary of essential information.

Total activity methods. The use—and misuse—of 'total activity' methods for alpha, beta, and gamma measurement have been discussed in detail (Preston *et al.*, 1972). There is little need for total-gamma assays within the FRL monitoring programmes since it is so easy to do full gamma-spectrometric analysis. In the case of alpha activity there is little demand for total counting and there are few significant discharges into the environment; consequently few monitoring programmes include monitoring of alpha-active radionuclides. In contrast, however, there is currently a particularly large demand from research, a reflection of the concern being shown about discharges from Windscale. Where alpha activity can be detected a separation procedure is usually adopted. In contrast, total-beta counting fulfils a useful role and most samples are analysed in this way, though usually only as a preliminary stage to further analysis (Dutton, 1968).

Gamma emitters. There are two types of counting system in routine use at FRL (Dutton *et al.*, 1974) differing according to the detector used. The older system uses a thallium-activated sodium iodide crystal, NaI(Tl); the newer development employs lithium-drifted germanium, Ge(Li), detectors. Each is associated with a multi-channel pulse-height analyser and makes use of the characteristic energy spectra of gamma-emitting radionuclides for detection and measurement; in consequence it is possible to resolve complex mixtures of radionuclides electronically without recourse to separative chemistry and so to greatly facilitate analysis of environmental samples. In its power to resolve complex mixtures the Ge(Li) detector is far superior to the Na(Tl) crystal but suffers from having to be maintained at constant low temperature ($-196\,^\circ$C). In practice, therefore, both systems are useful and both continue to be employed.

Preparative analysis of the samples prior to counting of the radioactivity in them can be kept to a minimum when using gamma spectrometry, and in many cases only physical procedures are necessary—drying and homogenization prior to packing into standard geometrical configurations for presentation to the detector. However, the measurement of gamma-emitting radionuclides in seawater will normally need extraction by chemical means, a particular example being the analysis of caesium-134 and -137 (Baker, 1975).

Alpha emitters. As with analysis of gamma emitters, use is made of the characteristic spectra which individual alpha emitters display and with the advent of high resolution silicon surface barrier layer detectors the need for separative radiochemistry has decreased. However, a chemical reaction is needed to free the activity from the inactive substrate material with which it is associated in the environment and it is usually necessary to separate the two main elements for which monitoring is needed, namely plutonium and americium because of spectral interference.

Beta emitters. Unlike alpha and gamma emitters, those which emit only beta activity cannot be individually analysed in mixtures by spectrometry and each must be separated for specific assay. It is for this reason more than any other that so much use is made at FRL of the total beta counting technique, which only requires the kind of preparation needed for estimation of gamma emitters. It is of no use for hazard assessment of mixtures of beta emitters and is used only to provide a rapid screening method to indicate significant change and to indicate whether recourse to more tedious separative analytical techniques is needed.

Total beta activity is measured using a thin-source, thin-window gas flow counter technique (Dutton, 1968). Separated samples are also counted in this way but use is also made of liquid scintillation counting and this is essential for tritium. Where a beta emitter also emits gamma radiation, assay using the latter is often preferred although beta counting may still be needed to attain the highest possible sensitivity of detection. Particular examples of beta emitters which have to be measured after separative chemistry are phosphorus-32, sulphur-35, strontium-90, technetium-99, and promethium-147.

Monitoring programmes in practice

A full account of the FRL monitoring programmes is published annually in the Technical Report Series 'Radioactivity in Surface and Coastal Waters of the British Isles' (Mitchell, 1977 and 1978). Reference to these reports is recommended to supplement the material quoted here which is used only to illustrate some of the marine monitoring programmes as they have worked out in practice, and which has been chosen to show what is required for a number of different kinds of source encompassing the whole nuclear power programme—reactor research and development, fuel fabrication, its use in nuclear power stations and the reprocessing that follows.

Reactor research and development. The UKAEA establishment at *Winfrith*, Dorset, has been responsible for several reactor development programmes, including latterly that on the Steam Generating Heavy

Water Reactor system; a complex mixture of radionuclides is released into the English Channel as a result. The critical pathway is through consumption of shellfish, particularly lobster and crab, but indicators are an important feature of the FRL routine monitoring programme with sampling of *Fucus* seaweed and limpets. Traces of some activation products are found in these indicators and characteristically they provide a very sensitive means of detecting the very low levels of radio-activity in this sector of the marine environment. They have the further advantage of being available at all seasons of the year. The operators of sites from which releases of radioactive waste are made have their own monitoring programmes, carried out in close collaboration with FRL as a result of the Ministry's role in the statutory control arrange-ments which normally include monitoring as a condition of the authorization. For Winfrith discharges, the UKAEA have their own monitoring programme (Flew, 1977), separate from that of FRL but complementary to it and together they provide a comprehensive coverage. Results of typical FRL measurements are shown in table 1.

Table 1   Radioactivity in indicator materials from the vicinity of Winfrith, 1976.

		Mean concentration of radioactivity[a]					
		Total beta		^{60}Co		^{65}Zn	
Material	**Sampling site**	Bq/kg	pCi/g	Bq/kg	pCi/g	Bq/kg	pCi/g
Oyster flesh	Poole Harbour	100	2.8	7	0.2	170	4.5
Crab flesh	Lulworth	130	3.4	41	1.1	70	1.8
Fish flesh	Weymouth Bay	120	3.3	< 4	< 0.1	< 4	< 0.1
Limpet flesh	Chapman's Pool	150	4.0	67	1.8	33	0.9
	Osmington Mills	100	2.8	22	0.6	7	0.2
Fucus serratus	Chapman's Pool	360	9.8	300	8.2	ND	ND
	Osmington Mills	370	10	140	3.7	ND	ND
	Weymouth	350	9.5	130	3.5	ND	ND
	Swanage	410	11	250	6.7	ND	ND
	Portland	370	10	70	1.9	ND	ND

[a]Specific activities referred to each material in the natural wet state.
ND This radionuclide was below the limit of detection in *Fucus serratus*.

Public radiation exposure is very low, less than one per cent of the ICRP-recommended dose limit in 1976.

Fuel fabrication. This is the main role of the plant at *Springfields*, Lancashire, from which low-level radioactive waste is released to the tidal Ribble estuary. The waste is composed of uranium residues and its daughter products and only a small amount of monitoring is needed, there being no internal pathways of possible significance but one in-volving external exposure. This is due to contamination of the mud of the river banks and the critical group are river authority workers who maintain these areas, particularly against flood damage. The essential monitoring required is measurement of the gamma radiation dose-rate but samples of the mud are also analysed. This reveals the presence of protactinium-234 m from Springfields, and also other radionuclides originating elsewhere (e.g. Windscale), and provides a good example of situations where there are coincident effects of more than one source. Despite this, public radiation exposure is very low and was less than one per cent of the ICRP-recommended dose limit in 1976 (see table 2).

Table 2   Radioactivity in mud and gamma dose-rates over the mud banks in the Ribble Estuary, 1976

| Sampling site | Mean concentration of radioactivity[a] | | | | | | | | Gamma dose rate | |
| | ^{106}Ru | | ^{134}Cs | | ^{137}Cs | | ^{234m}Pa | | | |
	Bq/kg	pCi/g	Bq/kg	pCi/g	Bq/kg	pCi/g	Bq/kg	pCi/g	nGy/h	μR/h
Pipeline outlet	2000	54	520	14	3200	87	29 000	790	350	40
Upstream										
90 m	2300	61	590	16	3800	104	39 000	1060	370	43
460 m	2500	68	590	16	4300	117	50 000	1350	370	43
Downstream										
90 m	2800	77	520	14	4600	124	33 000	880	350	40

[a]Specific activities referred to mud in the dry state.

In this case there are no particular needs for indicator materials for Springfields discharges because the effect of the discharges can be detected in the critical material.

Nuclear power stations. Bradwell, Essex, provides an interesting example of an estuarine monitoring programme, the particular feature to note being oyster consumption as the critical pathway. A number of radionuclides have been detected in oysters (Preston, 1968), all of them activation products except caesium-134 and -137. Currently the critical radionuclides are zinc-65 and silver-110 m, though the concentrations are now extremely low, as is the maximum dose-rate to members of the public, estimated 0.06 per cent of the ICRP-recommended dose limit in 1976. Samples of oysters are analysed regularly and, as representing the potential external exposure pathway, gamma dose-rate is measured along the foreshore in the intertidal zone. The values found are not significantly different from natural background and this is confirmed by analysis of samples of surface sediment by gamma spectrometry. To complete the surveillance local seaweed is collected as an indicator. Table 3 summarizes monitoring done in 1976.

Table 3   Radioactivity in environmental materials around Bradwell nuclear power station in 1976

	Mean concentration of radioactivity[a]										
	Total beta		^{137}Cs		^{65}Zn		^{110m}Ag		^{134}Cs		% of DWL
Material	Bq/kg	pCi/g	Bq/kg	pCi/g	Bq/kg	pCi/g	Bq/kg	pCi/g	Bq/kg	pCi/g	
Oyster	110	3.0	4	0.1	22	0.6	4	0.1	ND	ND	0.06
Fucus vesiculosus	230	6.3	7	0.2	ND	ND	<4	<0.1	ND	ND	—
Mud	1000	26.8	85	2.3	ND	ND	4	0.1	15	0.4	—

Gamma dose rate over intertidal mud = 63 nGy/h (7.2 µR/h)

[a]Specific activities referred to oyster and *Fucus vesiculosus* in the natural wet state, except mud which was dry.
ND These radionuclides were below limits of detection.

Sizewell, Suffolk, is quoted as a typical open coastal location for a nuclear power station. Local resources include both fish and shellfish and their consumption is the critical pathway for internal exposure. The critical pathway for external exposure is the familiar one of beach-occupancy but, as with fish/shellfish consumption, public radiation exposure is negligible and no artificial radioactivity or radiation exceeding detection limits can be found which are attributable to Sizewell operation. The trace of caesium-137 is consistent with fallout from nuclear weapons testing. In a rather barren coastline, where the beaches are almost entirely shingle, there are no convenient indicators close by.

With now more than a decade of experience in monitoring the effect of discharges at Sizewell the monitoring programme has been progressively reduced and very little is now needed. Samples of local fish and shellfish are analysed and gamma radiation dose-rate measured over the intertidal zone. Table 4 shows mean values of the data from surveys in 1976.

Table 4 Radioactivity in environmental materials around Sizewell nuclear power station in 1976

Material	Mean concentration of radioactivity[a]				
	Total beta		^{137}Cs		
	Bq/kg	pCi/g	Bq/kg	pCi/g	% of DWL
Round fish	120	3.2	4	0.1 }	< 0.1
Flatfish	200	5.5	4	0.1 }	
Lobster	92	2.5	< 4	< 0.1	< 0.01
Gamma dose rate over intertidal sand = 34 nGy/h 3.9 µR/h)					

[a]Specific activities referred to each material in the natural wet state.

Fuel reprocessing. Windscale, Cumbria, provides the most detailed example of a marine monitoring programme for discharges from a fuel reprocessing plant, not just within the UK but worldwide. The two critical pathways are internal exposure from consumption of fish and shellfish and external exposure from time spent on the foreshore, notably in the Ravenglass Estuary where the highest gamma radiation dose-rates are found. There are, of course, others through which lower levels of exposure occur; until the early 1970s consumption of laverbread made from the seaweed *Porphyra* was especially important. A great deal of current interest has been focused on resuspension of beach sediment and exposure caused by its inhalation. Evidence for this as a significant pathway is hard to find; current studies suggest that it may exist in some isolated areas but doses are lower than the maxima sustained from external gamma radiation. This pathway is presently being studied on a research basis and this example shows how monitoring and research can sometimes be difficult to distinguish.

The whole programme for Windscale is so large that only examples can be quoted here, taken from the fuller account published as an FRL Technical Report (Mitchell, 1977). Many samples of fish or shellfish of a variety of species typical of those which are caught commercially are collected and analysed, the predominant radionuclides in edible parts being caesium-134 and -137. Strontium-90, several plutonium radionuclides, and americium-241 are present but at lower concentration and much lower radiological significance than caesium-134 and -137; some analyses are made, though on a more selective basis than for caesium-134 and -137. This is partly because of their lower radiological significance, partly because concentrations can be inferred in these, and indeed sub-critical materials, with adequate accuracy from the discharge data. Detailed information on releases and their isotopic composition is provided by the operators of the plant, British Nuclear Fuels Limited, and a combination of past FRL research and monitoring experience on the distribution of each important radionuclide in the Irish Sea and its concentration factors in marine materials means that individual concentrations in these materials can be calculated as necessary. This shows the value of past experience and how it can be put to good use to make the most of available resources; however, it should not be taken to suggest that no additional measurements are made. Selected samples of fish and shellfish are analysed for some of those radionuclides of lower importance, a research programme on the distribution of radioactivity in the Irish Sea and beyond is maintained, presently concentrating on caesium-134 and -137 and the transuranics, and indicator seaweeds and molluscs are sampled.

A geographically extensive programme of monitoring the external exposure pathway is maintained even though long experience shows that the highest levels are generally to be found in the Ravenglass Estuary. Other areas where fine mud and silt collects show relatively high levels though values decrease with increasing distance from Windscale; doses over the intertidal zone of the open coast are substantially less than in estuaries because of the different nature of the sediment— coarse sand. The basic monitoring programme consists of regular surveys of the gamma radiation dose-rate supplemented by analyses of samples of sediment. Sediment is the most important compartment of the marine environment for a number of important fission products and transuranic elements, and the monitoring is closely integrated with a research programme which is being carried out on the behaviour of radionuclides on sediment; this seeks to answer questions about such issues as isotopic remobilization and sediment transport.

Doses through pathways from the discharge of waste from fuel reprocessing are higher than those from other sources: maximum individual values estimated for 1976 amounted to 8 per cent of the ICRP-recommended dose limit through the critical external exposure pathway, and to 44 per cent for the critical internal exposure pathway. In the latter case, a pessimistic basis is assumed such that the true maximum value is lower, as shown by whole body counting data generated during the Windscale Inquiry (Parker, 1978). Both of these maximum values relate to only a very few people and for the vast majority of the public doses are less, often very much less.

A representative selection of the analyses and measurements in 1976

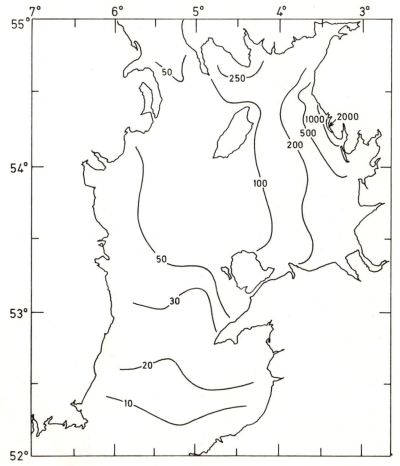

Figure 1 Distribution of caesium-137 in seawater in the Irish Sea, January 1976 (pCi 1⁻¹).

Table 5. Selected marine monitoring measurements in the vicinity of Windscale 1976

Material	Sampling point	Total beta Bq/kg	Total beta pCi/g	$^{95}Zr/^{95}Nb$ Bq/kg	$^{95}Zr/^{95}Nb$ pCi/g	^{106}Ru Bq/kg	^{106}Ru pCi/g	^{134}Cs Bq/kg	^{134}Cs pCi/g	^{137}Cs Bq/kg	^{137}Cs pCi/g	^{144}Ce Bq/kg	^{144}Ce pCi/g	^{239}Pu Bq/kg	^{239}Pu pCi/g	^{241}Am Bq/kg	^{241}Am pCi/g
Plaice	Windscale area (within 5 km of release point)	1 700	46	ND	ND	ND	ND	200	5.5	1 500	41	ND	ND	0.16	0.0042	0.26	0.0071
Cod		1 600	42	ND	ND	ND	ND	180	4.9	1 300	35	ND	ND	0.004	0.0001	0.02	0.0005
Crab		2 300	62	ND	ND	960	26	120	3.2	780	21	ND	ND	NA	NA	NA	NA
Mussel		13 000	360	700	20	9 600	260	150	4.0	850	23	670	18	110	3.1	340	9.2
Porphyra	Braystones	11 000	290	480	13	9 600	260	92	2.5	440	12	270	7.4	110	2.9	96	2.6
Mud	Ravenglass[b] Estuary (Newbiggin)	150 000	4100	3800	103	34 000	910	1400	39	11 000	290	12 000	330	4100	111	4800	130

Mean concentration of radioactivity[a]

[a]Specific activities referred to each material in the natural wet state, except mud which was dry.
[b]Mean gamma dose rate 1300 nGy/h (150 μR/h).
ND This radionuclide was below the limit of detection.
NA The samples were not analysed for these radionuclides.

is shown in table 5. Figure 1 illustrates some of the complementary research into distribution of radioactivity in the Irish Sea, in this case contours of caesium-137 from a survey by a research vessel in January 1976.

Conclusions

Marine environmental monitoring in the UK is essentially source-related and, designed and organized along critical path lines, provides a comprehensive surveillance system of the effects of operation of a large nuclear power programme. Monitoring is closely integrated with research programmes which also feed into the overall control system and so ensure that maximum information is gained from the effort available.

References

Anon (1959) *The Control of Radioactive Wastes*. Cmnd 884. London: Her Majesty's Stationery Office.

Baker, C. W. (1975) *The determination of radiocaesium in sea and fresh waters*. Technical Report 16. Lowestoft: Ministry of Agriculture, Fisheries and Food, Fisheries Laboratory.

Bryant, P. M. (1971) The derivation and application of limits and reference levels for environmental radioactivity in the United Kingdom. In *Health Physics Aspects of Nuclear Facility Siting*, ed. Voille) Voillequé, P. G. Vol. III, pp. 634–643. Idaho Falls, Idaho: The Health Physics Society.

Dutton, J. W. R. (1968) *Gross beta counting of environmental materials*. Technical Report FRL 3. Lowestoft: Ministry of Agriculture, Fisheries and Food, Fisheries Radiobiological Laboratory.

Dutton, J. W. R., Mitchell, N. T., Reynolds, E., and Woolner, L. E. (1974) Analytical systems applied to monitoring the aquatic environment in control of radioactive waste disposal. In *Environmental Surveillance of Nuclear Installations*, pp. 155–167. Vienna: International Atomic Energy Agency.

Flew, E. M. (1977) *Radioactive waste disposal by UKAEA establishments in 1976 and associated environmental monitoring results*. Report AERE-R8779. Harwell: United Kingdom Atomic Energy Authority.

IAEA (1976) *The Effects of Ionising Radiation on Aquatic Organisms and Ecosystems*. Technical Report Series 172. Vienna: International Atomic Energy Agency.

ICRP (1959) *Report of Committee II on Permissible Dose for Internal Radiation*. ICRP Publication 2. Oxford: Pergamon Press.

ICRP (1968) *Report of Committee IV on Evaluation of Radiation Doses to Body Tissues from Internal Contamination due to Occupational Exposure*. ICRP Publication 10. Oxford: Pergamon Press.

ICRP (1977) Recommendations of the International Commission on Radiological Protection. *Annals of the ICRP*, **1**. (ICRP Publication 26). Oxford: Pergamon Press.

Jefferies, D. F., Mitchell, N. T., and Hetherington, J. A. (1977) Collective population radiation exposure from waste disposal from a fuel reprocessing plant. In *Proceedings of the Fourth International Congress of the International Radiation Protection Association*, **3**, 929–932. Paris: The International Radiation Protection Association.

Mitchell, N. T. (1977) *Radioactivity in Surface and Coastal Waters of the British Isles 1976. Part 1: The Irish Sea and its Environs*. Technical Report FRL 13. Lowestoft: Ministry of Agriculture, Fisheries and Food, Fisheries Radiobiological Laboratory.

Mitchell, N. T. (1978) *Radioactivity in Surface and Coastal Waters of the British Isles 1976. Part 2: Areas other than the Irish Sea and its Environs.* Technical Report FRL 14. Lowestoft: Ministry of Agriculture, Fisheries and Food, Fisheries Radiobiological Laboratory.

Parker, The Hon. Mr Justice (1978) *The Windscale Inquiry*, Volume 1. London: Her Majesty's Stationery Office.

Pentreath, R. J. (1976) The monitoring of radionuclides. In *Biological Accumulators*, pp. 9–26. Rome: Food and Agriculture Organization of the United Nations.

Preston, A. (1966) The concentration of ^{65}Zn in the flesh of oysters related to the discharge of cooling pond effluents from the CEGB nuclear power station at Bradwell-on-Sea, Essex. In *Radioecological Concentration Processes*, ed. Aberg, B and Hungate, F. P. pp. 995–1004. Oxford: Pergamon Press.

Preston, A. (1968) The control of radioactive pollution in a North Sea oyster fishery. *Helgoländer wissenschaften Meeresunters*, **17**, 269–279.

Preston, A. (1971) The United Kingdom approach to the application of ICRP standards to the controlled disposal of radioactive waste resulting from nuclear power programmes. In *Environmental Aspects of Nuclear Power Stations*, pp. 147–157. Vienna: International Atomic Energy Agency.

Preston, A. and Jefferies, D. F. (1967) The assessment of the principal public radiation exposure from, and the resulting control of, discharges of aqueous radioactive waste from the United Kingdom Atomic Energy Authority factory at Windscale, Cumberland. *Health Physics*, **13**, 477–485.

Preston, A. and Mitchell, N. T. (1973) The evaluation of public radiation exposure from the controlled marine disposal of radioactive waste (with special reference to the UK). In *Interaction of Radioactive Contaminants with Constituents of the Marine Environment*, pp. 575–593. Vienna: International Atomic Energy Agency.

Preston, A., Fukai, R., Volchok, H. L., Yamagata, N., and Dutton, J. W. R. (1972) Radioactivity. In *A Guide to Marine Pollution*, ed. Goldberg, E. G. Ch. 7. New York: Gordon and Breach.

Monitoring the effects of domestic and industrial wastes

A. J. NEWTON, A. R. HENDERSON, and P. J. HOLMES

Clyde River Purification Board, East Kilbride, Glasgow, Scotland

Introduction

The marine environment supports a wide variety of human activities including fishing and mariculture, commerce and transport, mineral exploitation, recreation, and waste disposal. All these activities have clear biological implications; however, it is on the coastal ecosystems adjacent to major centres of population that the impact of such activities, particularly waste disposal, is most pronounced. Since the Industrial Revolution the volume and complexity of domestic and trade effluents entering inshore waters has reached significant proportions, and numerous estuaries and other nearshore areas have suffered serious pollution as a result. Today's Technological Revolution has made many demands on the ingenuity of the industrial chemist, and this is reflected in the composition of trade wastes which contain an ever-expanding range of exotic and often highly polluting substances.

In order to prevent, reduce, or at least contain pollution by domestic and industrial effluents, the statutory authorities have had to embark on programmes of monitoring and research so that water quality criteria can be defined and the maximum permissible levels of pollutants established.

Nature of domestic and industrial effluents

Domestic effluent may consist of 'domestic sewage'—household waste-water from residential areas—which has a reasonably uniform and well documented composition. More commonly, domestic effluent consists of 'industrial sewage' which is derived from mixed residential and industrial areas. Its composition is less predictable, depending to a large extent upon that of the industrial components, but it is usually sufficiently innocuous for it to be accepted, with domestic sewage, into the regional sewerage system and discharged to fresh or tidal waters after receiving treatment if necessary. Domestic sewage exerts a heavy bio-chemical oxygen demand (BOD), contains large numbers of faecal

171

bacteria and high concentrations of nutrients and suspended solids. As industrial sewage, domestic effluent may also contain elevated levels of heavy metals and persistent organic compounds.

As its name implies, industrial effluent is the wastewater from industrial plant, and often contains a small admixture of sewage from the workforce. Its composition will vary according to the manufacturing processes involved; thus it may be innocuous or highly toxic, even to the extent that it cannot be accepted into a sewerage system but must be disposed of separately after appropriate treatment. The parameters of interest in the effluent may again include BOD, nutrients, suspended solids, and metals, to which may be added oils and other complex hydrocarbons, extremes of pH, heat, etc.

Some effects of domestic and industrial wastes

The effects of wastes entering the coastal system will depend upon many factors, notably the composition of the waste itself and the physical, chemical, and biological characteristics of the receiving area. Some possible effects are detailed below.

1. Effluent exerting a high BOD may utilize dissolved oxygen (DO) in the receiving waters to the extent that DO tensions fall too low to support a normal biota. In extreme circumstances, all the DO may be consumed and anaerobic conditions established; hydrogen sulphide and methane gases are then evolved which may give rise to smell nuisance; widespread mortalities may occur among aquatic organisms and the water may prove corrosive to the condenser tubes of ships and power stations.

2. Elevated concentrations of nutrients may promote prolific growths of planktonic and attached algae, which will, in due course, exert a BOD. Sometimes the phytoplankton blooms consist of 'undesirable' species which may be directly toxic or unpalatable to herbivores. Sometimes also, large accumulations of seaweed are cast ashore on the beaches where they decompose anaerobically and provide a smell nuisance.

3. Heated effluent, such as cooling water, may subject the biota to thermal shock, perhaps with lethal effects, or reduce the oxygen-carrying capacity of the water.

4. Heavy metals, PCBs, pesticides, and other conservative substances may exert direct toxic effects if present in sufficient concentration. Some organisms selectively accumulate these pollutants, even when the latter are present in the environment in only minute quantities, and, by concentration via the food web, high order carnivores can incur significant and even lethal body burdens.

5. Domestic effluent may seriously contaminate beaches and nearshore

waters, rendering them unattractive, and the levels of faecal bacteria present may make them unfit for bathing or growing shellfish according to forthcoming and proposed EEC legislation; they may also render existing shellfish stocks unfit for human consumption.

6. Radioactive wastes, which may enter the coastal environment from military installations, hospitals, power stations, etc., and petroleum wastes may also have serious effects, but these categories of pollutant have been described already by other authors (see this volume).

Along developed coastlines, most discharges are made to waters already contaminated to a greater or lesser degree by domestic and industrial effluents. Interactions between these may provide a complex, seemingly bewildering, array of effects.

Role and organization of monitoring

The major role of monitoring in marine pollution prevention has been described by the Royal Commission on Environmental Pollution (1971, 1972), the Central Unit on Environmental Pollution (Department of the Environment, 1974), and by the recently formed Marine Pollution Monitoring Management Group (Department of the Environment, 1977). Details of the legislative and administrative organization of pollution control have been given by CUEP (Department of the Environment, 1976), and the MPMMG have described the pattern of commitment and responsibility for monitoring at regional, national, and international levels.

At the national level, the principal statutory authorities involved in monitoring are the Scottish River Purification Boards, the Regional Water Authorities, the Ministry of Agriculture, Fisheries and Food, the Department of Agriculture and Fisheries for Scotland, and the Ministry of Agriculture for Northern Ireland. Although the United Kingdom was considered to be probably one of the most intensely monitored of industrial countries, it was accepted that the value of this effort was severely lessened by sweeping differences in the monitoring aims and methods of the different authorities. Accordingly, the MPMMG was given the task of assessing the needs, formulating the strategy, and coordinating the efforts of the various authorities. The Group drafted a strategic plan for an idealized national monitoring programme based on four objectives which were, in decreasing order of priority:

1. To provide information on current levels of contaminants of fish and shellfish from a human consumption standpoint.

2. To provide information required to form an adequate knowledge base on which the risk to resources could be established.

3. To establish current levels and subsequently to maintain under surveillance trends in pollutant concentrations.

4. To establish a system of biological monitoring (Department of the Environment, 1977).

Monitoring domestic and industrial wastes in the Clyde

The Clyde River Purification Board is responsible for controlling pollution from effluent discharges from half of Scotland's population and industry, taking in an area of some 14 000 km^2 and a long coastline which will, incidentally, be tripled in length following the implementation of Part II of the Control of Pollution Act 1974 within the next two years.

Pollution control is effected by a licensing system, whereby a discharge is permitted only if specific conditions relating to the composition and quantity of the effluent, and the position of the outfall, are strictly complied with. Definition of the appropriate conditions for each discharge can pose an almost intractable problem, and would be quite impossible without information, gained by monitoring and research, on:

   (a) the identity of the pollutant sources,
   (b) the character, loading and concentration of the pollutants, and
   (c) the mode of entry, behaviour, biological effects, and ultimate fate of the pollutants.

This information may then be used to
   (*i*) identify polluted or vulnerable areas, following comparison with unpolluted areas;
   (*ii*) establish whether the degree of pollution is static, worsening or improving;
   (*iii*) assess the need for, and requirements of, controls;
   (*iv*) judge the effectiveness of controls already in force—relaxing or tightening these as may be appropriate;
   (*v*) evaluate the capacity of the system to accept additional polluting loads;
   (*vi*) follow the restoration of, or departure from, satisfactory conditions;
   (*vii*) judge the suitability of the area for various forms of usage, such as bathing or rearing shellfish; and
   (*viii*) give early warning of trouble ahead so that any approaching threat of pollution can be detected and forestalled (Newton, 1978b).

Because of its large area, and the multiplicity of polluting domestic and industrial discharges, the Board has sought to develop a coastal monitoring scheme which is swift, effective, and efficient (Newton, 1978b). The scheme now in use employs the following three categories of parameters.

Physico-chemical parameters

The status of the area as a whole is assessed using physico-chemical parameters because these are fairly simple and cheap to measure and can be expressed in unequivocal terms. There are many examples to choose from, including non-conservative determinands such as DO levels which may be significant in the local context (see Mackay and Leatherland, 1976), to more conservative substances posing a hazard on a much wider scale. These substances may include many of those mentioned in List I and List II of the EEC Council directive on pollution caused by dangerous substances discharged into the Community's aquatic environment (EEC, 1976b).

List I comprises the following groups of substances selected mainly on the basis of their toxicity, persistence, and bioaccumulation:

(*a*) organohalogens,

(*b*) organophosphorus,

(*c*) organotin compounds,

(*d*) substances which are carcinogenic in the aquatic environment,

(*e*) mercury and its compounds,

(*f*) cadmium and its compounds,

(*g*) persistent mineral oils and hydrocarbons of petroleum origin, and

(*h*) persistent synthetic substances, which may interfere with any use of the waters.

List II of the directive includes:

(*a*) substances belonging to the families and group of substances in List I for which no limit values have been determined;

(*b*) (*i*) the following metals, metalloids and their compounds: Zn, Cu, Ni, Cr, Pb, Se, As, Sb, Mo, Ti, Sn, Ba, Be, B, U, V, Co, Tl, Te, Ag;

(*ii*) biocides not appearing in List I;

(*iii*) substances liable to cause tainting of products for human consumption from the aquatic environment;

(*iv*) toxic or persistent compounds of silicon;

(*v*) phosphorus and its inorganic compounds;

(*vi*) non-persistent mineral oils and hydrocarbons of petroleum origin;

(*vii*) cyanides and fluorides;

(*viii*) substances such as NH_3 which have an adverse effect on the oxygen balance.

The appropriate selection can be made on the basis of presence in the effluent, the uses to which the area is, or could be, put, etc. There are of course dangers in focusing attention on the levels of individual determinands:

(*a*) one is more interested in the biological effects of a given level of pollutant 'X' than in that level of pollutant 'X' itself, and
(*b*) no account is taken of possible synergistic or antagonistic inter-action between the determinands themselves and the host of other substances in the coastal environment (Newton, 1978b).

Aesthetic and microbiological parameters

Aesthetic quality is usually evaluated in terms of the appearance, smell, and sometimes the taste of the subject. Unfortunately, the quantifica-tion of the sensory stimuli received is very subjective and will vary between individual observers according to their range of experience and expectations of the environment (Newton, 1978b). The Board's efforts towards monitoring the aesthetic effects of polluting discharges using a modified form of the Garber (1960) method of receiving water analysis have, however, met with qualified success.

Aspects of the quality of beaches and nearshore waters, and indeed the performance of sewage outfalls, have been evaluated successfully by monitoring the degree of contamination by faecal micro-organisms (Hammerton, 1978). Microbiological and physico-chemical parameters form the basis of the definition of the EEC Commission's quality re-quirements for bathing water (1976a) and for waters favourable to shellfish growth (1976c), and may be used to monitor the effects of effluents with regard to these activities.

Biological parameters

Although complicated by difficulties in interpretation, the monitoring of the effects of domestic and industrial wastes according to biological parameters is the most meaningful and potentially the most sensitive. The biota react to the integrated bio-physico-chemical environment, in-cluding incoming pollutants, and they may well be sensitive to pollu-tants at concentrations which may be difficult to determine by conven-tional analytical means. A basic requirement of monitoring is that the criteria used should be simple to measure, be sensitive, and be able to detect and identify subtle pollution effects which may destroy coastal ecosystems over the long term. 'Biological indicators' can, potentially at least, meet this requirement (Newton, 1978b).

The assessment of biological quality has usually been based on the presence or absence, distribution, and abundance of organisms at the population or community level. This approach is illustrated by con-tinuing studies of (*a*) the return of benthic macroinvertebrates to the severely polluted Clyde Estuary (Newton, 1978a, 1978b; Mackay, Tayler, and Henderson, 1978) and (*b*) the diversity of the macrobenthos in Irvine Bay, an area in the Firth of Clyde which receives heavy loadings of domestic and industrial waste (Clyde River Purification Board, 1977). Such studies are of special value when assessing long-term trends or specifying environmental objectives, such as the two simple biological criteria that the Royal Commission on Environmental Pollution (1972) specified for the management of estuarine waters. Unfortunately, biological assemblages tend to respond only slowly to the effects of environmental change, and population estimates thus fail to meet the 'swift response' requirement of pollution monitoring. Also, sample analysis can be very time-consuming and demands considerable skill in identification.

A great many other aspects of populations have been considered, such as departure from the 'normal' numerical relationship between different trophic groups, or between different meiofaunal taxa such as harpacticoid copepods and nematodes. The success of all such studies is, however, prejudiced by the difficulty one has in distinguishing, with confidence, between natural and pollution-induced changes (Newton, 1978b).

Assessment and monitoring at the individual or species level of organization seems very promising from the practical viewpoint, bearing in mind the need for speed, simplicity, and sensitivity. One approach involves the identification and use of 'indicator species', albeit with caution (see Eagle and Rees, 1973). Such species have been described by Glover (in Departments of Environment and Transport, 1978) as analogous to the miner's canary; as such they afford an *ad hoc* test of significant change rather than serving as long-term monitors of ecological well-being or change. Another approach, which offers considerable potential for both local and global scale monitoring, involves the collection of ubiquitous 'pollutant-accumulators' such as mussels and certain seaweeds and measurement of their pollutant content (see Bryan, 1978). In making the selection, consideration must be given to the behavioural adaptations of the species (Davenport, 1977), and much work has yet to be undertaken to deduce the biological implications of the levels of pollutant recorded. A third approach involves the use of bioassay and stress-indexing, both in the field and the laboratory, to express the potency of an effluent or water sample in terms of physiological, biochemical, or cytological criteria (see Bayne, 1978). The latter approach is perhaps best suited to regional-scale monitoring, but it requires further development before it can be used by the pollution control authorities.

The Board presently monitors the levels of pollutants in the rivers, incoming effluents, and the waters, sediments, and biota of the coastal environment. It monitors the effects of domestic and industrial wastes in terms of physico-chemical, aesthetic, and microbiological and biological parameters—the latter including whole population, 'indicator species', and 'pollution-accumulator' approaches. Considerable scope remains for the development of improved monitoring methods; the usefulness of those we have is severely limited by procedural and interpretational difficulties.

Hammerton (1978) has highlighted the paradoxical situation which has arisen during the current period of economic difficulty—on the one hand the pollution control authorities have been urged to cut expenditure and accept a slow down in the progress of environmental recovery; whereas, on the other hand, the same authorities are being called upon to undertake monitoring additional to their own requirements. Unfortunately, the majority of the statutory authorities will be unable to participate in coordinated monitoring as fully and as rapidly as they would wish.

Acknowledgements

We thank Mr D. Hammerton, our Director, for his permission to present this paper and acknowledge the major part played by our colleagues, notably Dr Leatherland, Messrs Tayler, Jickells, Mostyn, and Frame, and the crew of the S.V. *Endrick II* in our coastal monitoring work.

References

Bayne, B. L. (1978) Stress and pollution research at IMER. In *Biological Indicators of Estuarine Pollution—Research and Application*, pp. 21–28. Departments of the Environment and Transport, Research Report No. 22. London: HMSO.

Bryan, G. W. (1978) Research at MBA on heavy metal contamination in estuaries. In *Biological Indicators of Estuarine Pollution—Research and Application*, pp. 11–17. Departments of the Environment and Transport, Research Report No. 22. London: HMSO.

Clyde River Purification Board (1977) *Water Quality, a Baseline Report*. East Kilbride: Clyde River Purification Board.

Davenport, J. (1977) A study of the effects of copper applied continuously and discontinuously to specimens of *Mytilus edulis* (L.) exposed to steady and fluctuating salinity levels. *Journal of the Marine Biological Association of the United Kingdom*, **57**, 63–74.

Department of the Environment (1974) Monitoring of the environment in the United Kingdom. *CUEP Report*. London: HMSO.

Department of the Environment (1976) Pollution control in Great Britain: how it works. *CUEP Pollution Paper No. 9*. London: HMSO.

Department of the Environment (1977) Monitoring the marine environment of the United Kingdom. *CUEP Pollution Report No. 2*. London: HMSO.

Departments of the Environment and Transport (1978) *Biological Indicators of Estuarine Pollution—Research and Application*. Research Report No. 22. London: HMSO.

Eagle, R. A. and Rees, E. I. S. (1973) Indicator species—a case for caution. *Marine Pollution Bulletin*, **4**, 25.

EEC (1976a) Council directive concerning the quality of bathing waters. *Official Journal of the European Communities*, No. L31, 5.2.76.

EEC (1976b) Council directive on pollution caused by certain dangerous substances discharged into the aquatic environment of the Community. *Official Journal of the European Communities*, No. L129, 18.5.76.

EEC (1976c) Proposed council directive relating to the quality requirements for waters favourable to shellfish growth. *Official Journal of the European Communities*, No. COM(76), 570.

Garber, W. F. (1960) In *Proceedings of the First International Conference on Waste Disposal in the Marine Environment*, ed. Pearson, E. A. pp. 372–403. New York: Pergamon.

Hammerton, D. (1978) EEC directives on the quality of bathing water and on water pollution caused by the discharge of dangerous substances—the River Purification Board viewpoint. In *Report of Symposium on River Pollution Prevention, Ingliston, 15th March, 1978*, pp. 13–28. Institute of Water Pollution Control.

Mackay, D. W. and Leatherland, T. M. (1976) In *Estuarine Chemistry*, ed. Burton, J. D. and Liss, P. S. pp. 185–218. London: Academic Press.

Mackay, D. W., Tayler, W. K., and Henderson, A. R. (1978) The recovery of the polluted Clyde Estuary. *Proceedings of the Royal Society of Edinburgh*, **76B**, 135–152.

Newton, A. J. (1978a) Macrobenthos as indicators of quality changes in the Clyde Estuary. In *Biological Indicators of Estuarine Pollution—Research and Application*, pp. 43–53. Departments of the Environment and Transport, Research Report No. 22. London: HMSO.

Newton, A. J. (1978b) Coastal quality control and assessment. In *Coastal Pollution Control, 3, Danish International Development Agency, Jutland*, pp. 917–929.

Royal Commission on Environmental Pollution (1971) First report. London: HMSO.

Royal Commission on Environmental Pollution (1972) Third report. *Pollution in some British estuaries and coastal waters*. London: HMSO.

Marine wildlife conservation in the coastal zone

ROGER MITCHELL

*Chief Scientist Team, Nature Conservancy Council,
Godwin House, Huntingdon, Cambs*

Introduction

The necessity to develop a strategy for marine wildlife conservation in British coastal waters is in large part a reaction to the many conflicting activities in this zone which seem to grow both in type and magnitude, and in the area which they affect. Who, for example, in the last decade, could have foreseen the exploitation of the North Sea oil fields and predicted the enormous and rapid growth of oil handling facilities and related industries in remote coastal areas? Many other impacts have been in existence for a long time, but have increased in scale in recent years and, in consequence of the deleterious changes brought about, are now regarded as serious threats to the ecological integrity of certain coastal areas. Impacts in this category include land reclamation, establishment of coastal structures, effluent discharges, coastal dumping, shipping accidents, mineral extraction, and fisheries. Activities like recreation, educational and scientific collecting, and the introduction of alien species, while often being localized, none the less cause considerable disturbance to ecosystems in many sectors of the coastal zone. In addition there are a number of new uses being considered which have the potential for inflicting perhaps greater modification of the marine environment than the present established uses. Proposals for the development of new water resource schemes may affect areas directly by the construction of reservoirs on intertidal flats, while the manipulation of river flow will alter the freshwater flows to estuaries and coastal waters both quantitatively and qualitatively, affecting biological processes. The search for alternative forms of energy has led to proposals for tidal barrages and wave energy generators which might cause permanent changes in the tidal regime and exposure affecting adjacent coasts and coastal waters. Related to this is the proposal that the harvesting of kelp on a vast scale could provide for the production of gas in commercial quantities by the anaerobic digestion of the plant material.

The purpose of this essay is to describe the basis of a possible

181

strategy for marine wildlife conservation in the littoral and shallow sublittoral areas of Britain's coastal zone. Survey, surveillance, and monitoring are discussed briefly in relation to this strategy.

Strategy for marine conservation

In the terrestrial environment the need for nature conservation stemmed from the recognition that human impact in Britain was so extensive that hardly a habitat or species was safe from the direct or indirect effects of development unless deliberate protective measures were taken. Although much of the marine environment bordering Great Britain is still in a fairly natural state compared with the land, the sea has by no means escaped man's modifying influence. Mitchell (1977 and in press) discussed the actual and potential impacts on estuaries and the shallow seas and concluded that it was necessary to effect a conservation strategy while there was yet time. Further deleterious change might not only affect areas that are currently of conservation interest, but could also result in the fragmentation of the remaining interest. The parallel here is seen in lowland Britain where areas of recognized conservation importance often occur as small islands in a sea of urban, industrial, and agricultural development. Any delay in the implementation of a site conservation policy might result in a reduction in the availability of suitable conservation areas, one consequence of which could be the limiting of opportunity to find sites in a suitable location and of the right size where conflicts of use could be minimized.

In developing a strategy for marine conservation it has seemed logical to adopt, as far as is practical, the philosophy and practice that has evolved on the land. While a superficial examination of the requirements for conservation in the marine environment may appear to present many basic differences, in fact familiarity with the terrestrial approach emphasizes the similarities. In addition, it is desirable that procedures adopted for marine conservation harmonise with those accepted for the land as a broadly comparable overall approach will ensure continuity and efficiency of operational practices.

The body charged with the task of ensuring that the objectives of nature conservation are met in Britain is the Nature Conservancy Council (NCC), established by Act of Parliament in 1973 as successor to the former Nature Conservancy. It has two primary and complementary functions—species/site safeguard and advice supported by an appropriate research programme. The principal strategy on land has really been the conservation of habitat, partly by the establishment of a National Nature Reserve (NNR) series and partly as a function of the advisory role.

Habitat conservation

National Nature Reserves. These statutory reserves are owned, leased, or managed by agreement with the owner. Establishment of such reserves is a continuous process and ideally every nationally important site should be an NNR. Under present legislation, the NCC can declare NNRs only down as far as the mean low water mark of ordinary tides in England and Wales and the mean low water mark of spring tides in Scotland. However, it now seems appropriate to extend NCC's remit at least into the shallow sublittoral. Following recent representations from NCC, the Department of the Environment has established an interdepartmental working party to examine the need for new legislation and the form that legislation might take.

The principal purposes of nature reserves as set out in 1947 in Cmd 7122 (Ministry of Town and Country Planning, 1947) are entirely appropriate for the requirements of marine reserves:
1. *Conservation and maintenance.* To preserve and maintain as part of the nation's natural heritage places which can be regarded as reservoirs for the main types of community and kinds of wild plants and animals represented in this country, both common and rare, typical and unusual, as well as places which contain physical features of special or outstanding interest.
2. *Survey and research.* To provide areas where both fundamental and applied scientific research and survey, whether of a long-term or a short-term nature, can be carried out by members of the proposed national Biological Service[1], of research institutions and university departments, and by amateurs, in the certain knowledge that these reserves will be managed and maintained in such a way as to enable work of this kind to be planned and carried out undisturbed by considerations other than those that arise from the scientific controls which it will be essential to impose on such work if the reserves are to be conserved and maintained for the fulfillment of functions as in (1).
3. *Experiment.* To provide reserves exclusively for experimental purposes.
4. *Education.* To provide educational facilities insofar as these are compatible with purposes (1) and (2).
5. *Amenities.* To provide places where the nature-lover—whether layman or amateur student—can go to enjoy nature.

Sites of Special Scientific Interest. A key element in NCC's advisory function in safeguarding sites is achieved by the notification of Sites of Special Scientific Interest (SSSI). Under the National Parks and Access to the Countryside Act 1949, the NCC has a duty to notify local

[1] This proposal resulted in the Nature Conservancy being established by Royal Charter in 1949.

planning authorities of any land, not already managed as a nature reserve, which is of special interest because of its biological, geological, or physiographical features. Details of these SSSIs are submitted to local authorities in the form of county schedules and maps. SSSIs fulfil an important nature conservation function both regionally and nationally. Their contribution as part of the national heritage has been recognized in the context of fiscal policy by their eligibility in certain cases for exemption from Capital Transfer Tax. Many SSSIs are of national importance and are likely to become NNRs in due course. NCC is empowered under the Countryside Act 1968 (as amended by the Nature Conservancy Council Act 1973) to enter into agreements with the owners of SSSIs to provide financial support for any management practices which contribute towards nature conservation aims on these sites. Although SSSIs are not protected to the same degree as are the NNRs, the Town and Country Planning General Development Order 1977 does provide that planning authorities must consult NCC before granting permission for the development of land in an SSSI. This consultation allows NCC to assess the impact of the proposed development and, where possible, advise on how any damage might be reduced to an acceptable level.

As is the case with NNRs, NCC's duty to notify SSSIs is limited to the land above the low water mark, although NCC's general advisory role has no such limitation. Indeed NCC is frequently consulted on potentially damaging developments below the low water mark such as gravel extraction, dredging, routes of pipelines, the fixing or mooring of large structures, and effluent disposal. NCC has also advised on the establishment of a number of voluntary marine reserves.

Species conservation

Complementary to habitat conservation is the protection of species, wherever they occur, as under the Protection of Birds Acts and the Conservation of Wild Creatures and Wild Plants Act 1975. However, it seems as if the legal protection of a species may not be so useful in the sea as on land. There are two difficulties: one is the determination of what factors are causing the species to be rare, and the other is obtaining the evidence that the species is endangered. Apart from commercial species, the organisms at greatest risk are those large, attractive species which have considerable appeal to the amateur collector. But even in this case, over-collecting would probably lead only to local scarcity. Much more information is required on the distribution, abundance, and ecology of marine species before the need and form of species conservation appropriate to marine organisms can be properly assessed. In fact, there is no real alternative to habitat conservation in the protection of the smaller, inconspicuous species.

Selection of marine conservation areas

Although habitat conservation by the establishment of NNRs and SSSIs was developed for terrestrial ecosystems, there seems no reason why these concepts (albeit in a modified form, could not work equally well in the sea. Indeed, even without new legislation, a great deal still remains to be achieved in establishing the sites of conservation importance in the littoral zone. Until very recently the only site attributes that were considered when assessing these marine sites were higher plants and ornithological interest. While this has led to the inclusion of a number of sedimentary shores of conservation interest in the NNR and SSSI series, rocky shores are usually included only incidentally in these sites, and the sublittoral is represented only by the bed of drainage channels in estuaries.

In discussing the selection of marine conservation areas, it will be useful to consider further the approach already used for terrestrial sites, its relevance to the marine environment being for the most part quite plain. The purpose of the SSSI series is to ensure that there is at least a basic minimum of notified sites which include enough habitats to support viable populations of all species in Great Britain. Such a basic minimum should also include a sufficient geographical spread of sites to ensure the present distribution of ecosystems and species. This is particularly important in the long term because many species are likely to change their distribution at some time as a result of climatic or, possibly, genetic change. This system also protects a greater range of sites than the NNR series, which they support by providing second-best sites if the best sites are lost.

The 'key site' concept

The best and most important of the SSSIs are known as 'key sites', and these have been chosen to represent all the main types of natural and semi-natural vegetation with their characteristic communities of plants and animals. In *A Nature Conservation Review* (NCR), Ratcliffe (1977) has explained the 'key site' concept and the methods by which these sites are assessed and selected. It should be stressed that not all SSSIs are key sites, and not all key sites are NNRs, although ideally they should be. Of the 652 SSSIs that occur on the coast, 273 are key sites, but only 39 are NNRs.

The Nature Conservation Review strategy

The basic NCR strategy of site assessment, as outlined below, provides a very useful framework for the future task of selecting those littoral and sublittoral sites that should be afforded some protection. This process is basically the same whether one is considering just the key sites or all sites of conservation importance. Of course, just as there are modifications in approach and emphasis in selecting sites within different terrestrial formations, so will these be developed to take account of differences in the marine environment. In addition, the strategy is being continually refined as the process of key site selection and the notification of SSSIs and the designation of NNRs continues.

In selecting a series of sites which gives acceptable representation of all the more important features within the range of variation of ecosystems in Britain, two main steps are involved:
1. The identification and recording of primary scientific data for the site in terms of environmental and biological characteristics.
2. On the basis of the intrinsic site features, and within the framework of a classification of ecosystem types, to assess comparative site quality and select those sites which should constitute the national series of key conservation areas.

The NCC has already made a start in drawing together information on littoral and sublittoral sites in Great Britain and, where necessary, is initiating new surveys. Amateur groups and voluntary bodies are also giving more attention to these areas and have established their own surveys and projects, providing much additional, valuable information. On the basis of the initial data, a classification of littoral and sublittoral sites is being developed appropriate to the site selection process. However, the qualitative knowledge of the range of sublittoral ecosystems is not yet an adequate basis for the comparison and assessment of different sites of similar quality. Nevertheless, sufficient information exists for intertidal ecosystems and the process of selecting littoral sites of conservation importance has now started.

Criteria for comparative site assessment. By established practice ten criteria have become accepted by which the nature conservation value of terrestrial sites can be judged. The relevance of these criteria for the selection of nature conservation areas in the marine environment is discussed below.
1. *Extent.* In general, the bigger the area, the better, providing other attributes are equal. Below a certain minimum size the communities or species to be conserved may be adversely affected by adjacent activities —'the edge effect'. There is, therefore, a certain minimum size necessary to ensure the integrity of the site—'the viable unit' concept. However, the optimum size of site will vary according to the type of formation concerned. In considering marine sites the problem will, in some cases, be one of restricting the choice of area to a reasonable size. Nevertheless, there will be instances where localized physical conditions will produce

distinctive sites of limited extent, such as tidal rapids, brackish lagoons, and marine caves.

In considering the size of area necessary to conserve a particular representative type of marine community, the vagaries of the recruitment of certain species will need to be taken into account. Apart from the temporal instability that may exist, there is also the spatial element. Thus the area chosen should ideally be large enough so that fluctuations in the occurrence of species due to differential settlement of the planktonic stages are so far as is practical encompassed by the area selected.

2. *Diversity*. Variety in terms of species and communities is highly desirable, and depends to a large extent on the physical diversity of an area and the number of different communities it supports. However, diversity may also be influenced by other factors, for example, a rock face, sheltered from the waves and tides will generally support a greater variety of species than an adjacent exposed rocky headland. Community types which are of intrinsic low diversity are no less interesting or important than types with a characteristically high diversity; species richness should be treated as a relative and not absolute factor in this context.

3. *Naturalness*. In general, an area which is unmodified by human influence is desirable. However, on the land this is a difficult criterion to apply, for in many cases a high conservation value is placed upon entirely artificial habitats. If 'naturalness' can be equated with 'living natural resources' then a definition of the latter which may be useful is simply 'any species or ecosystem that is not intensively farmed', this description being accepted by the International Union of Nature and Natural Resources at the launch of the World Conservation Strategy in 1978. Thus, while the marine environment may be considered to be in a fairly natural state, areas with high degrees of 'naturalness' being abundant, sites that are remote from potential disturbance or exploitation are nevertheless valuable.

4. *Rarity*. While rarity on a national scale has been the grounds for the establishment of 'species reserves' by voluntary conservation bodies, the Nature Conservancy Council's policy has generally been to regard rare species as a bonus on sites selected for other reasons. Rarity of species may reflect rarity of habitat, or be due to an organism occurring at the limits of its distribution in the biogeographical sense. Species with good dispersal mechanisms may be rare because their settlement and establishment is unpredictable, particularly in the marine environment. It is therefore necessary to understand what factors are operating to make a species rare before it is given weight in an evaluation exercise or its management needs are defined. However, other things being equal, the presence of rare species on a site gives it a higher value than another comparable site that has no rarities.

5. *Fragility.* This criterion reflects the sensitivity of habitats, communities, and species to environmental change. While all natural and semi-natural habitats are susceptible to human impact, some are more intrinsically sensitive than others, and in extreme cases their viability may be doubtful. However, this criterion would seem to have more application on land than in the marine environment, except perhaps in the case of certain rare habitats that might easily be altered through pollution or physical destruction.

6. *Representativeness.* It is not only necessary to choose areas which are in some way unusual or unique, but it is also desirable to represent the typical and ordinary sites which contain habitats, communities, and species which occur commonly or are widespread. Such sites may be particularly important as experimental areas where, for example, homogeneity may be a desirable feature in designing experiments which require many replicate or reference plots. In the marine environment there is a relatively wide choice of representative sites in a natural condition, so that selection might be almost arbitrary.

7. *Recorded history.* The extent to which a site has already been used for scientific study and research is a factor of some importance, and will elevate the value of such a site above another similar site which has little recorded history. Certainly the value ascribed to marine areas which are the location of long-term studies and experiments, and indeed the classic 'collecting sites', is enhanced by the application of this criterion.

8. *Position in an ecological/geographical unit.* Where practical it is desirable to include within a single geographic area as many as possible of the important and characteristic ecosystems, communities, and species of that region. Thus, where two comparable sites are assessed, and one is contiguous with another site representing a different formation or ecosystem, then this site is regarded as of higher value. This criterion is obviously related to those of size and diversity. There is also the practical convenience of a composite site for wardening and management. For example, it would be convenient to have marine sites adjacent to terrestrial sites in some cases, especially where it might be desirable to control access.

9. *Potential value.* This criterion is concerned with the potential for the rehabilitation of sites or the re-creation *de novo* of an example of an ecosystem. While this concept has far more relevance to the terrestrial situation, one could conceive of a marine example where mud flats lost by dredging or reclamation were artificially reinstated by engineering works which encouraged accretion.

10. *Intrinsic appeal.* It is inevitable that certain ecosystems or organisms will attract more interest than others because of the bias in human interest. Attractive and popular groups are given more weight in

assessing a site than the more obscure groups. Thus epibenthic macrofauna will tend to be used in site assessment rather than the meiofauna, and similarly the macro-algae will be 'more significant' than micro-algae. Nevertheless it will be important to ensure that less popular groups of organisms are adequately represented in key sites.

Having applied the above criteria to compare and arrange in importance the respective merits of sites of similar character, there remains perhaps the more difficult task of choosing the national series of key sites. The complex procedure has been analysed by Ratcliffe (1977), who describes the integration of the appraisal of information and the application of criteria for site assessment as an almost automatic process, comparable to a person who, crossing a busy street, makes numerous observations and calculations about his own motion and those of the vehicles, without ever thinking consciously about the matter. Despite the difficulties, the NCR has listed and described 735 chosen key sites. It will necessarily be many years before such a comprehensive and soundly based review can be completed for littoral and sub-littoral ecosystems. Nevertheless, there is every hope that within a reasonable period it will be possible to produce a list of marine sites which are of sufficient conservation importance to be regarded as key sites and potentially suitable for designation as statutory marine reserves. Indeed, in 1947, thirty years before the present NCR was published, the Wildlife Conservation Special Committee in their report (Ministry of Town and Country Planning, 1947) proposed 73 sites which in their 'unanimous opinion' should be established as National Nature Reserves. The value of these sites has been upheld by subsequent investigations and, with the exception of those sites since degraded, many are included in the NCR.

The detection and assessment of change

It is very much a part of NCC's advisory role to comment on potential impacts on the marine environment and assess any expected or demonstrable change in terms of its implications for wildlife conservation. The detection of change depends on programmes of investigation involving 'survey', 'surveillance', or 'monitoring'. While the original meanings of these terms have long since been obscured by their scientific adoption, even their sense in scientific usage has been distorted by misapplication. Therefore, before discussing further the detection of change, it is necessary to re-define these terms.

Survey: An inventory of the attributes of a site in terms of the physical habitat and the associated organisms, in qualitative and quantitative terms, including seasonal data. A survey may be restricted to a particular species or group of species, the result being a description of what occurred where over a given timescale, e.g. the distribution and abundance of *Zostera* in an estuary.

Surveillance: A procedure by which a series of surveys is conducted in a sufficiently rigorous manner for changes in the attributes of a site to be detected and followed over a period of time. In this case, the standard against which change is measured is arbitrary and is usually the original or 'base-line' survey, e.g. changes in the distribution and abundance of *Mytilus edulis* on a rocky shore. Where a surveillance exercise is mounted with the purpose of identifying abnormal changes, it is necessary to establish a 'base-line' which shows the range of normal changes. In this case the standard can usually be defined only after a long series of surveys. This process where extraordinary changes are detected against a background of ordinary or normal fluctuations is often aptly described as separating the signal from the noise.

Monitoring: This is a more restricted term than 'surveillance' (though in usage it often encompasses it) and describes the measurement of change against a known or given standard by a series of rigorous surveys or tests, e.g. the levels of coliform bacteria in shellfish measured against public health standards.

There is really no substitute for basic survey in obtaining descriptive information for the classification, assessment, and selection of conservation areas. While a marine conservation strategy is still being established there is bound to be a high requirement for and a high priority given to site surveys. The data required to assess the effects of environmental change depend on impact studies supported by basic ecological research into structure, functioning, and regulation of marine communities. In this respect the research requirements for nature conservation purposes are really no different from the research programmes of many organizations carrying out or supporting marine research, particularly those effects related to pollution studies. However, studies on the detection, rather than the assessment, of change are rather a different matter.

Most of the surveillance and monitoring procedures used in the past, and many of the current ones, are based on estimates of abundance or the distribution of organisms, and depend on criteria related to the functioning of the population and the community rather than the individual species. These procedures are essential for ecological research and provide very necessary data for determining management programmes; in fact they are ideal for many nature conservation needs. However, population estimates are often historical records and are virtually useless when applied to the more immediate problems of practical surveillance and monitoring. For example, in the short term such estimates cannot be used to detect subtle and sub-lethal effects on organisms which, in the long term, might cause drastic changes in population levels. In addition, a numerical assessment technique can take no proper account of metabolic and genetic adaptation to change. In the practical world where one is often required to give instant advice on the implications for nature conservation of a particular scheme or development, one needs to have rapid answers on the effects of a particular impact on the environment. And these answers need to be accurate enough to provide sufficient information for one to assess whether the actual or potential change is acceptable. In some instances it may be too late to rely on next year's answer from numerical assessment procedures, and even then the change could be so obvious as to be picked up by a single survey or the field observations of one experienced investigator.

An alternative is to seek more immediate answers through investigating the physiological response of species to specific pollutants and other changes in the environment. Clues that a change is occurring may also be detected by the establishment of appropriate cytological and biochemical indices for affected species and the use of specially selected organisms for bioassay tests. Although these procedures are relatively new, it is essential that they be rapidly developed and widely adopted. This is not to say that methods used in surveillance at the population or community level and those used at the species or cellular level are mutually exclusive. Indeed, in very many instances they should be used side by side, if for no other reason than to determine the extrapolation between a response at the species level and its eventual result on the population or community. In this way the predictive and interpretative facilities are continually refined. This approach to monitoring and surveillance is also made very necessary by the United Kingdom's posture to pollution studies in defining standards according to the effects of pollutants.

The main concern of some operating in the customer/contractor area of pollution science is the magical properties attributed to what are often merely crude surveys. Allied to this is the whole business of interpreting change, assuming it can actually be demonstrated. Many so-called surveillance exercises carried out at present have no clear objectives. The level of change that is sought is not defined, nor is any attempt made to design a programme to collect the information whereby one might reasonably detect and quantify the cause of the change; without this surely the whole point of carrying out the exercise is lost. For example, how can a particular effluent or one of its component chemicals be identified as responsible for causing a detectable alteration in a particular population or ecosystem when no attempt whatsoever is made either to analyse the effluent or to research the interaction between the component chemicals and the organisms concerned? Unfortunately, many examples of such inadequate surveillance and monitoring programmes are in current use.

Perhaps it is asking too much to require that such elaborate and sophisticated techniques be used; it would certainly be very expensive. Also, many of the methods are still being researched and standardized, and have not reached the stage of practical application. A move in the right direction would be the achievement of a higher level of professional integrity; there is a real need for honest dealings both by those who are practitioners in the art of monitoring and those who are in receipt of these skills as customers. The word 'honest' is used advisedly, because there does seem to be a good deal of deliberate evasion or neglect of the real issues in this business. If this is not the result of dishonesty, then these procedures are still brought into disrepute, because the only other interpretation of an inadequate research programme is that the practitioners and their customers are not as discerning or as adequately briefed as they ought to be. Of course many customers and contractors go to some lengths to ensure that a monitoring programme is soundly based; these remarks are really directed at those who merely want to be seen to be monitoring, whatever the level of their responsibility is for a potential impact. In this case, science on a shoestring and public relations research is the all too familiar consequence.

Conclusion

The basic and inevitable conclusion from this paper is that enormous resources are needed both to fund all the tactical and strategic research required and to manage and maintain an adequate marine conservation strategy. Not only is there a high requirement for more survey, but a great deal more effort is required in the field of fundamental ecological research. More needs to be done to elevate the art of monitoring and surveillance, and this requires much more experimental data on the individual performance of organisms under specific types of stress, and how any changes at this level are manifested in the population and community. At least part of the resources requirement could be satisfied by establishing that a particular level of funding should be directed to appropriate research activities for each new development which has the potential to affect marine ecosystems. Finally, one needs to decide who should monitor the monitoring programmes, and what the standard should be.

Acknowledgement
The author wishes to thank Catherine Gosse for much useful discussion and for critically reading the manuscript.

The views expressed in this paper are for the most part those of the author, and do not necessarily indicate any commitment by the Nature Conservancy Council.

References

Ministry of Town and Country Planning (1947) *Conservation of Nature in England and Wales: Report of the Wildlife Conservation Special Committee (England and Wales).* (Cmd 7122) London: HMSO.

Mitchell, R. (1977) Marine Wildlife Conservation. In *Progress in Underwater Science 2 (new series),* ed. Hiscock, K. and Baume, A. D. pp. 65–81. London: Pentech Press.

Mitchell, R. (in press) Nature Conservation implications of hydraulic engineering schemes affecting British estuaries. *Hydrobiological Bulletin.*

Ratcliffe, D. A. ed. (1977) *A Nature Conservation Review.* Cambridge University Press.

Discussion

The account that follows is a resumé of the main points that arose in the four discussions held during the symposium. In some cases the general remarks made by the Chairmen of Sessions are also summarized.

The first session was devoted to intertidal areas. The Chairman (Holliday) remarked that surveys and monitoring form the basis of Britain's wildlife conservation programme on land, yet efforts in this direction in marine habitats are patchy to say the least. Many countries have marine nature reserves backed by effective legislation, and hopefully Britain will not be without such reserves for much longer. The time is ripe to delimit more sites of special scientific interest not only on the shores but also below low-tide and in the seas around this country. Perhaps marine biologists here, despite Britain's long record of scientific activity in this field, have not sufficiently emphasized the need for protection nor pressed for sufficient backing for monitoring programmes. The Nature Conservancy Council is giving increasing priority to marine and coastal sites. It would be aided in this endeavour if marine biologists who hold the view that a conservation programme for marine wildlife is important were to ensure that national agencies, and particularly the NCC, are aware of their views at the highest level. R. Mitchell (NCC) said that there are at present nearly 700 sites notified as of special scientific interest round our coasts, but there is still a need to refine the selection of prime sites including those holding rare marine species whose case may not have been championed sufficiently strongly hitherto.

An ever-present problem with nature reserves is that of effective protection. Even in the case of terrestrial reserves it is difficult enough to police a site against the damaging activities of visitors or even of site owners in some instances; the problem is more acute in marine reserves, especially when they are in remote areas. The NCC is all too aware of these problems, and sees the need for encouragement to be given to watchful observers, plus, in large areas, an effective wardening system. In several areas of Britain, such as Skomer, Pembrokeshire, and Purbeck, voluntary 'policing' is organized, and this year on Lundy in the Bristol Channel a seasonal warden was on hand to oversee the activities of visiting marine naturalists and divers.

A recurring point raised by discussants concerned the scale of changes taking place on a shore that could reasonably be monitored. An important decision that has to be made early in a monitoring survey is

195

whether the sort of programme that is practicable will yield the necessary information and detect the appropriate scale of change. The Anglesey team (Jones *et al.*), for instance, were interested in natural changes, and to detect these their methods involved visiting sites at a frequency that could be maintained and that could reasonably answer most of the likely questions asked of the resulting data. So of their 30 sites, 12 were visited monthly, and the rest either three-monthly, half-yearly, or yearly. Nelson-Smith pointed out the difference between those surveys that seek to detect natural change and those that answer an immediate need, such as the imminent approach of an oil slick. The carefully designed base-line or long-term study may be more accurate, but the need for rapid action may preclude laying on a survey that will answer all questions asked of the data when the oil has come ashore. In some cases it may be necessary to utilize the help of non-professional observers or those still under training; such assistance can be a highly important part of the survey, provided that the instructions issued are clear-cut and standardized.

The point was made by Mitchell that in the future more attention should be given to tests such as those detecting a physiological response of organisms to pollutants. All too seldom is an effective test investigated, and indeed research councils would be likely to welcome an opportunity to back well-designed investigations into the use of physiological indicators or the use of single or grouped organisms as monitors. Physical and chemical parameters are being measured alongside the biological, usually by teams other than the biologists, but a major problem here is that their results are too often contracted to water development concerns or industry, with publication restrictions imposed. Yet in the end, however successful the monitoring, and however widespread its net and fine its mesh, the decisions on acceptability of the situation revealed will rest with an authority that may be essentially lay. So an important corollary to any monitoring programme must be that its results are understandable to those who will make the policies. This point was returned to in the final session of the symposium.

The second session of the symposium, on the sublittoral and below, began with a paper by W. F. Farnham, E. B. G. Jones, N. A. Jephson, and P. W. G. Gray entitled 'Eulittoral and sublittoral algae: monitoring and distribution of species introduced into the Solent' which is not published here. Farnham traced the reasons for monitoring studies by the team from Portsmouth. The programme was initiated mainly because of the unusual combination of circumstances present in the central southern waters of Britain, including the fact that the area is a focal point for the chance introduction of exotic species, perhaps by the agency of shipping together with the effects of commercial and industrial activity. The group's background work was well justified when in 1973 the Japanese sea-weed *Sargassum muticum* first appeared in the area, possibly brought in with the commercial oyster, *Crassostrea gigas*. The weed is now spreading from its original centre in the Solent, and there is concern that if it becomes established in the clear waters of the South-West it may grow even more prolifically there. The point emerged from this example that early recognition of introduced species is critical if one is to trace consequent changes in the invaded ecosystem.

Submarine monitoring is still an underdeveloped field. Hiscock listed many techniques available to the SCUBA diver that are appropriate for monitoring work, but added that many of them have yet to be properly evaluated. Another problem is the variability of individual observers and the reliability or otherwise of relating one set of results to others. This becomes particularly important when the survey team is composed of amateurs, which, just as in shore work, can potentially extend the capability of monitoring to a geographical coverage and a time-scale that is hardly ever possible using only professionals. Earll (Projects Coordinator, Underwater Conservation Programme) remarked that Underwater Conservation Year 1977 and its successor, UCP, had shown how valuable an asset to biological studies the amateur diving fraternity is, and how well able such people are to tackle biological projects. At least 400 divers in the United Kingdom had taken some part in projects during UCY 77, and as a result experiments and surveys had been conducted that would otherwise have been impossible.

Although there are some experiments that can be done now only with deep submersibles, Rowe did not think that these vessels would be widely used in general monitoring work. For one thing, they are expensive to operate, in both money and time; for another, they are feared by deep-sea animals and may not therefore give the opportunity for behavioural or numerical study that has sometimes been envisaged for them. By the same token, underwater houses and laboratories, while providing excellent opportunity for the study of human psychology in the alien world of the sea, cannot yet be seen as particularly useful tools for monitoring studies. More useful information is likely to emerge in the immediate future from the use of remote devices, including physico-chemical sensors, photography, and television.

The third session took as its running theme the Open Ocean. The Chairman (Fogg) referred to the man-induced hazards that urgently require monitoring, and emphasized the importance of continual vigilance by all available means. He remarked that DDT has been detected in oceanic water at a concentration as high as 0.2 in 10^{11}, which is a fifth of the level that is lethal to the brine-shrimp. The discussion of Parry's paper revealed the possible conflict between chemical and mutagenic effects: only when the effects of the one are clearly delimited can the effects of the other be properly assessed. For instance, now that the chemical nature and effects of polyaromatic hydrocarbons are beginning to be known, the possible mutagenicity of this important range of pollutants can be assessed.

Colebrook had pointed out in his paper that the plankton monitoring programme has been in progress for about 30 years, the longest-running monitoring survey to date. Significantly, the design of the survey at its inception has remained substantially unchanged, so that earlier results can be compared meaningfully with results obtained now. A principal feature emerging is the cyclical nature of planktonic occurrences, and though these are being correlated with climatic and other factors, a serious shortcoming of the whole survey is the lack of physico-chemical data taken alongside the biological readings: not even the basic nutrients are recorded at the same time as the plankton samples are taken. The cyclical phenomena are the more interesting since they are regarded as wholly natural, plankton being the one group of organisms that is not significantly affected yet by man's activities. Herring (Institute of Oceanographic Sciences) commented that the IOS is involved in work on the deep-sea nekton, etc., for which almost no base-line information is available, so that for this enormous part of the marine environment it is almost impossible to detect change.

A major point that arose during Laws' paper on monitoring whale and seal populations, and in the subsequent discussion, was the state of our knowledge of the krill populations, on which not only the marine animals but also a separate industry now depends. Laws was asked whether we have sufficient knowledge of the life history, standing stock, and production of krill to say whether an industry exploiting it would affect the whale population. He replied that it is clear that the level of krill available to the surviving whales and to other consumers in south polar waters has already been affected by reduction of the whale stocks. What is required now is a total management plan for the Southern Ocean, a plan that takes note of natural competition between endemic populations. The Antarctic Treaty Powers are currently discussing the draft of a Convention for the Conservation of Antarctic Marine Living Resources which will, it is hoped, lead to the establishment of a management regime—before a large krill industry develops. Our knowledge of marine mammal species and in particular the differences between separate breeding populations within species, is being added to by such techniques as blood typing, gel electrophoresis, and biometric comparisons. It is becoming clear that there is only slight interchange between separate breeding stocks.

The final session was devoted to the effects on the marine environment of human activity. The Chairman (Hammerton) said that the Stockholm Conference of 1973 had provided a great impetus to environmental studies by governments and one that had arisen was the Global Environmental Monitoring System (GEMS) which formed part of the Earthwatch Programme of the newly created UNEP. In Britain the Harmonised Monitoring System, comprising 189 stations on rivers, is now providing an input to GEMS and the Marine Pollution Monitoring Management Group is endeavouring to provide the framework for a properly coordinated programme to monitor our coastal waters. However, he feared that in the present economic climate such monitoring was likely to remain decidedly patchy for some years to come. Newton (Clyde River Purification Board) remarked that statutory authorities concerned with environmental quality were very conscious of the need to monitor, but they had been urged to reduce expenditure and to accept a slow-down in the rate of improvement, while, at the same time, undertaking monitoring that was additional to their own immediate needs.

The final stages of the discussion ranged across the general theme of the whole symposium. Legislation has not yet come into force to bring about monitoring of the marine environment at the level thought desirable. Some participants felt that Government coordination was necessary. It will be brought into effect when the Control of Pollution Act 1974 is fully implemented in 1979. But this Act will not determine the level of monitoring, only the statutory requirement to carry it out, which some authorities are well on the way to fulfilling anyway. What is required now is a firm directive to water authorities and other statutory bodies, as they assume responsibilities beyond the mouths of rivers, to coordinate their activities in the direction of marine monitoring to fit in with what is required in both national and international waters, as determined by Central Government. One immediate step that can be taken (and will be taken under the 1974 Act) is to remove the shackles of confidentiality that now prevent free publication of results. Ultimately, whatever our monitoring programmes discover and whatever the actions that their results suggest, the community at large must find the money to bring about the required improvement. Biologists must set their house in order, principally in deciding what degrees of change they can realistically measure, and in relation to this, what levels of change they regard as environmentally unacceptable. Their 'miner's canary' must be reliable and acted upon. Their views and the reasons for them must be understandable to the body politic, for the decisions that Governments make will be based on their advice.

Index

References to information in figures and tables are given in italics.